INTRODUCTION TO AXIOMATIC SET THEORY

SYNTHESE LIBRARY

MONOGRAPHS ON EPISTEMOLOGY,

LOGIC, METHODOLOGY, PHILOSOPHY OF SCIENCE,

SOCIOLOGY OF SCIENCE AND OF KNOWLEDGE,

AND ON THE MATHEMATICAL METHODS OF

SOCIAL AND BEHAVIORAL SCIENCES

Editors:

DONALD DAVIDSON, *Rockefeller University and Princeton University*

JAAKKO HINTIKKA, *Academy of Finland and Stanford University*

GABRIËL NUCHELMANS, *University of Leyden*

WESLEY C. SALMON, *Indiana University*

THÉORIE AXIOMATIQUE DES ENSEMBLES
First published by Presses Universitaires de France, Paris
Translated from the French by David Miller

Library of Congress Catalog Card Number 71–146965

ISBN 90 277 0169 5

Printed in The Netherlands by D. Reidel, Dordrecht

JEAN-LOUIS KRIVINE

INTRODUCTION TO AXIOMATIC SET THEORY

D. REIDEL PUBLISHING COMPANY / DORDRECHT-HOLLAND

INTRODUCTION

This book presents the classic relative consistency proofs in set theory that are obtained by the device of 'inner models'. Three examples of such models are investigated in Chapters VI, VII, and VIII; the most important of these, the class of *constructible sets,* leads to Gödel's result that the axiom of choice and the continuum hypothesis are consistent with the rest of set theory [1][1].

The text thus constitutes an introduction to the results of P. Cohen concerning the independence of these axioms [2], and to many other relative consistency proofs obtained later by Cohen's methods.

Chapters I and II introduce the axioms of set theory, and develop such parts of the theory as are indispensable for every relative consistency proof; the method of recursive definition on the ordinals being an important case in point. Although, more or less deliberately, no proofs have been omitted, the development here will be found to require of the reader a certain facility in naive set theory and in the axiomatic method, such as should be achieved, for example, in first year graduate work (*2ᵉ cycle de mathématiques*).

The background knowledge supposed in logic is no more advanced; taken as understood are such elementary ideas of first-order predicate logic as *prenex normal form*, *model of a system of axioms*, and so on. They first come into play in Chapter IV; and though, there too, all the proofs (bar that of the reduction of an arbitrary formula to prenex normal form) are carried out, the treatment is probably too condensed for a reader previously unacquainted with the subject.

Several leading ideas from model theory, not themselves used in this book, would nevertheless make the understanding of it simpler; for example, the distinction between intuitive natural numbers and the natural numbers of the universe, or between what we call formulas and what we call expressions, are easier to grasp if something is known about

[1] Numbers in brackets refer to items of the Bibliography, which is to be found on p. 98.

non-standard models for Peano arithmetic. Similarly, the remarks on p. 44 will be better understood by someone who knows the completeness theorem for predicate calculus.

All these ideas can be found for example in [4] (Chapters I, II, III) or [5] (Chapters I, II, III).

The approach of the book may appear a little odd to anyone who thinks that *axiomatic* set theory (as opposed to the naive theory, for which, perhaps, this is true) must be placed at the very beginning of mathematics. For the reader is by no means asked to forget that he has already learnt some mathematics; on the contrary we rely on the experience he has acquired from the study of axiomatic theories to offer him another one: the theory of binary relations which satisfy the Zermelo/Fraenkel axioms. As we progress, what distinguishes this particular theory from other axiomatic theories gradually emerges. For the concepts introduced naturally in the study of models of this theory are exactly parallel to the most fundamental mathematical concepts – *natural numbers, finite sets, denumerable sets*, and so on. And since standard mathematical vocabulary fails to provide two different names for each idea, we are obliged to use the everyday words also when referring to models of the Zermelo/Fraenkel axioms. The words are thereby used with a completely different meaning, the classic example of this being the 'Skolem paradox', which comes from the new sense that the word 'denumerable' takes when interpreted in a model of set theory.

Eventually it becomes obvious that even the everyday senses of these words are by no means clear to us, and that we can perhaps try to sharpen them with the new tools which are developed in the study of set theory. Had this problem been posed already at the beginning of the study it would have been tempting to shirk it, by saying that mathematics is merely the business of manipulating meaningless symbols.

It is an open question how much set theory can do in this field; but it seems likely that what it can do will be of interest beyond mathematical logic itself.

Note: This translation incorporates a number of additions and corrections to the original French edition.

CONTENTS

CHAPTER I

THE ZERMELO/FRAENKEL AXIOMS OF SET THEORY

When we formulate the axioms of set theory, we have to rely on an intuitive understanding of sets, in much the same way that we develop the axioms for a vector space from commonsense ideas about three-dimensional space. Despite this, once we have the axioms of set theory written down we are free to study any other structure in which they hold, just as we study many vector spaces over and above the Euclidean space R^3. Thus set theory is no different from any other axiomatic theory familiar to the reader. It is, like the theories of groups, rings, fields, vector spaces, lattices, and so on, an *abstract* theory.

A structure satisfying the axioms of set theory is called a *universe*. A universe $\mathscr{U}$ is basically a collection of objects called *sets*. We do not say 'a *set* of objects', since what we are going to call sets are just those objects of which $\mathscr{U}$ is a collection. It is a straightforward precaution when defining vector spaces, for example, not to use the same word both for the vector space and for a vector in it. The same applies here.

Also involved in $\mathscr{U}$ is a single undefined binary relation, a relation which links some of the objects of $\mathscr{U}$ with others. It is called the *membership relation*, and is denoted by $\in$. We read the expression '$x \in y$' as 'x belongs to y', or 'the set x belongs to the set y', or 'x is a member of y', or 'x is an element of y', or even 'the set y contains the element x'.

Since $\in$ is a relation on the universe $\mathscr{U}$, it holds only between sets. The symbol '$\in$' and the words 'belong', 'contain', 'member', and 'element' will be used only when this particular relation is intended; and should we ever use any of these words in their everyday senses, we shall state this explicitly. For example, we will not normally say 'the object x is an element of the collection $\mathscr{U}$', but 'x is in $\mathscr{U}$'.

So a universe can be pictured as a graph of the kind shown in figure 1; the arrow indicates that the object at its head is an element of the one at its tail. For example, $b \in a$, $c \in c$.

Nothing said so far imposes any restriction on the binary relation $\in$. In order to do just this we now begin to list the axioms of set theory.

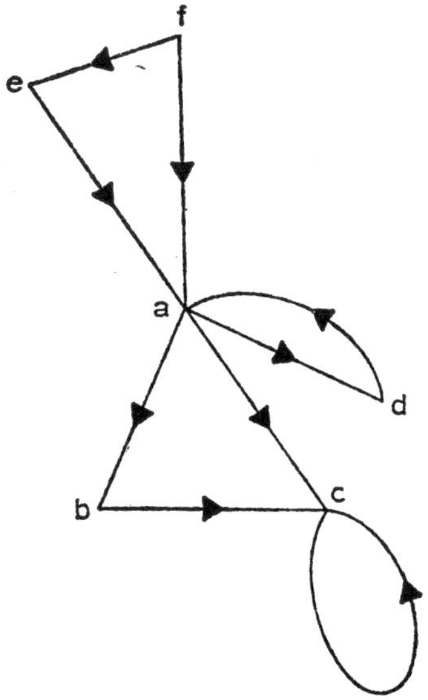

Fig. 1.

1. AXIOM OF EXTENSIONALITY

No two distinct sets in $\mathscr{U}$ have the same elements. We write this as

$$\forall x \forall y [\forall z (z \in x \leftrightarrow z \in y) \rightarrow x = y].$$

This axiom is not satisfied by the binary relation shown in the figure; for b and c are different, yet each has c as its only element.

A useful 'axiom' to put down now, though we will not number it since it turns out to be a consequence of later ones, is the *pairing axiom*. Given two sets a and b, this axiom guarantees the existence of a set c containing a and b as its only elements. By the axiom of extensionality there is only one such set c. We write the pairing axiom as

$$\forall x \forall y \exists z \forall t [t \in z \leftrightarrow t = x \vee t = y].$$

We write $\{a, b\}$ for the set c whose only elements are a and b. A special case is when a and b are the same set; it follows from the axiom that there is a set $\{a\}$ containing a, and a alone.

If $a \neq b$ we call $\{a, b\}$ a *pair*. The set $\{a\}$ is sometimes called a *singleton* or *unit set*.

From two sets a and b three applications of the pairing axiom yield the set $\{\{a\}, \{a, b\}\}$, called the *ordered pair* of a and b and usually denoted by $\langle a, b\rangle$. The following theorem justifies this definition.

THEOREM: *If* $\langle a, b\rangle = \langle a', b'\rangle$ *then* $a = a'$ *and* $b = b'$.
PROOF: If $a = b$ then $\langle a, b\rangle = \{\{a\}\}$, and so $\langle a, b\rangle$ has only one element. Thus $\langle a', b'\rangle$ is also a singleton, which means that $a' = b'$. It follows that $\{\{a\}\} = \{\{a'\}\}$, and so $a = a'$; thus $b = b'$ too.

When $a \neq b$, $\langle a, b\rangle$ is a pair, and so $\langle a', b'\rangle$ must also be one. This can only be true if $a' \neq b'$.

But as $\{\{a\}, \{a, b\}\} = \{\{a'\}, \{a', b'\}\}$, either

$$\{a\} = \{a', b'\} \quad \text{and} \quad \{a, b\} = \{a'\}$$

or

$$\{a\} = \{a'\} \quad \text{and} \quad \{a, b\} = \{a', b'\}.$$

The former possibility is ruled out simply because $\{a\}$ is a singleton whilst $\{a', b'\}$ is a pair. Thus $\{a\} = \{a'\}$ – and so $a = a'$ – and $\{a, b\} = \{a', b'\}$ – and so $b = b'$. ▮

If a, b, c are sets, the set $\langle a, \langle b, c\rangle\rangle$ is called the *ordered triple* of a, b, c, and is denoted by $\langle a, b, c\rangle$.

THEOREM: *If* $\langle a, b, c\rangle = \langle a', b', c'\rangle$ *then* $a = a', b = b', c = c'$.
PROOF: From the identity of $\langle a, \langle b, c\rangle\rangle$ and $\langle a', \langle b', c'\rangle\rangle$ we get $a = a'$ and $\langle b, c\rangle = \langle b', c'\rangle$. Thus also $b = b'$ and $c = c'$. ▮

In the same way we can define the ordered quadruple $\langle a, b, c, d\rangle$ by

$$\langle a, b, c, d\rangle = \langle a, \langle b, c, d\rangle\rangle,$$

and for any $n > 0$ the ordered n-tuple $\langle a_0, a_1, \ldots, a_{n-1}\rangle$ by

$$\langle a_0, a_1, \ldots, a_{n-1}\rangle = \langle a_0, \langle a_1, \ldots, a_{n-1}\rangle\rangle.$$

The previous theorems generalize.

THEOREM: *If* $\langle a_0, a_1, \ldots, a_{n-1}\rangle = \langle a_0', a_1', \ldots, a_{n-1}'\rangle$ *then* $a_0 = a_0', a_1 = a_1', \ldots, a_{n-1} = a_{n-1}'$.
PROOF: Obvious by induction on n. ▮

It should be noted that, given three distinct sets a, b, c in the universe $\mathscr{U}$, we do not yet have any way of proving the existence of a set d containing just a, b, and c as members. The next axiom fills this gap.

2. UNION AXIOM (OR SUM-SET AXIOM)

According to this axiom, to every set a there corresponds a set b whose members are precisely the members of the members of a. We write this formally as

$$\forall x \exists y \forall z [z \in y \leftrightarrow \exists t (t \in x \wedge z \in t)].$$

The set b, called the *union of the members* of a (or, more briefly, the *union* of a) and denoted by $\bigcup_{x \in a} x$ (or $\bigcup a$), is unique; for any set b' having the same property would necessarily have the same elements as, and so be identical with, b.

This axiom solves the problem posed above. For if a, b, and c are sets, the set $d = \bigcup \{\{a, b\}, \{c\}\}$ contains a, b, and c, and nothing more besides. We write the set d as $\{a, b, c\}$.

More generally, any *finite* number of sets $a_0, a_1, \ldots, a_{n-1}$ can be collected together into a set $\{a_0, a_1, \ldots, a_{n-1}\}$ containing them and them only. This is easily proved by induction on n; we need only observe that $\bigcup \{\{a_0, a_1, \ldots, a_{n-2}\}, \{a_{n-1}\}\}$ has the desired property for n if $\{a_0, a_1, \ldots, a_{n-2}\}$ has it for $n-1$.

If a and b are sets, then the union of $\{a, b\}$ is called the *union* of a and b, and written $a \cup b$. It is trivial that $a \cup (b \cup c) = (a \cup b) \cup c = (b \cup a) \cup c =$ = the union of the set $\{a, b, c\}$.

The same applies for any finite number of sets $a_0, a_1, \ldots, a_{n-1}$; the union of the members of $\{a_0, a_1, \ldots, a_{n-1}\}$ is called the union of a_0, a_1, and ..., and a_{n-1} and written $a_0 \cup a_1 \cup \ldots \cup a_{n-1}$.

3. POWER-SET AXIOM

Let a and b be sets in $\mathscr{U}$. The statement $\forall x (x \in a \rightarrow x \in b)$ is abbreviated $a \subset b$, and read 'a is *included* in b' or 'a is a *subset* of b'.

The power-set axiom tells us that, given a set a, there exists a set b whose members are just a's subsets; in symbols it becomes

$$\forall x \exists y \forall z [z \in y \leftrightarrow z \subset x].$$

By the axiom of extensionality only one set b can have this property. It is called the *power-set* of a and written $\mathscr{P}(a)$.

It is worth emphasizing at this point that we shall now use the words 'subset' and 'include' only in the sense given them above, to express a relation between two objects of the universe. This sense is quite different from the usual one (when we say, for instance, that a class is included in the universe). We call it the formal sense of the word 'include' (or 'subset'), and the usual one will be called the intuitive sense. To avoid from the start any possible confusion that might arise, let us agree that when we use the words 'subset' and 'include' without qualification, we are using them *in their formal senses*; any use of these words in their intuitive senses will be noted as such explicitly.

For instance: each set a defines a subset (in the intuitive sense) of the universe $\mathscr{U}$, namely the part A composed of the elements of a. Now if b is a set included in a, then the corresponding B is included (in the intuitive sense) in A. But there may be – and indeed often are – subsets (in the intuitive sense) of A to which no object of the universe corresponds; to which, that is, there is no corresponding subset of a.

Before giving the remaining axioms of set theory we must discuss how relations, over and above those we already have, can be defined on $\mathscr{U}$. At present we have only the membership relation $x \in y$, and the equality relation $x = y$ (which is satisfied by a and b if and only if a and b are the same object). The following rules allow us to define others.

I. Given a relation (of three arguments, for instance) $R(x, y, z)$, and an object a in $\mathscr{U}$, then we can define a binary relation $R(a, y, z)$, to be satisfied by the objects b and c iff $R(a, b, c)$ holds.

II. Given a relation (three arguments, say) $R(x, y, z)$, then we can define a binary relation $R(x, x, z)$; this is satisfied by a and b iff $R(a, a, b)$ holds.

III. Given any relation (say of two arguments) $R(x, y)$, we can define $\neg R(x, y)$, a relation satisfied by a and b iff $R(a, b)$ does not hold.

Likewise from two relations (of three arguments again) $R(x, y, z)$ and $S(u, x, v)$ we can define a relation $R(x, y, z) \vee S(u, x, v)$ of five arguments satisfied by a, b, c, d, e iff one or other of $R(a, b, c)$ and $S(d, a, e)$ holds.

IV. Again, from a ternary relation $R(x, y, z)$ we can define a binary relation $\exists y R(x, y, z)$ which is satisfied by a and c iff $R(a, b, c)$ holds for some b in $\mathscr{U}$.

Starting from the two binary relations $x \in y$ and $x = y$ we can, by repeated applications of these rules, obtain relations of any number of arguments.

Suppose that $R(x, y)$ and $S(y, z)$ are two binary relations; then the relation $\neg R \vee S$ is written $R \to S$. The relation $\neg(\neg R \vee \neg S)$ is written $R \wedge S$; it is satisfied by a, b, and c iff $R(a, b)$ and $S(b, c)$ both hold. Similarly, $R \leftrightarrow S$ abbreviates $(R \to S) \wedge (S \to R)$, and is satisfied by a, b, c iff $R(a, b)$ and $S(b, c)$ either both hold or both fail.

Likewise, the singulary relation $\neg \exists x \neg R(x, y)$, written $\forall x R(x, y)$ is satisfied by b iff $R(a, b)$ holds for every object a in $\mathscr{U}$.

Relations constructed from $x \in y$ and $x = y$ by rules I–IV are thus defined by formulas composed (though not in an arbitrary fashion) from the symbols $=, \in, \neg, \vee, \exists$, the variables $x, y, z, u, v, \ldots$, and objects in $\mathscr{U}$.

It is clear that further relations on $\mathscr{U}$ can be defined by other methods. But *we shall not consider them in this book.*

A relation $R(x)$ of a single argument is usually called a *class*. A class is a subcollection (intuitively speaking) of the universe $\mathscr{U}$. For example, the formula

$$\forall u [u \in x \to \exists v [v \in x \wedge \forall t (t \in v \leftrightarrow t = u \vee t \in u)]]$$

defines a class; the formula

$$u \in x \to \exists v [v \in x \wedge \forall t (t \in v \leftrightarrow t = u \vee t \in u)]$$

defines a binary relation $R(u, x)$. So if a is in $\mathscr{U}$, the formula $R(a, x)$, namely

$$a \in x \to \exists v [v \in x \wedge \forall t (t \in v \leftrightarrow t = a \vee t \in a)],$$

defines a class.

Objects of $\mathscr{U}$ which appear in a formula E are called the *parameters* of E. $R(u, x)$, for example, has no parameters; $R(a, x)$ has the object a as its sole parameter.

A formula of no arguments at all, a *closed formula* or *sentence* as it is called, is either true or false in the universe. For example, the formula $\exists x \exists y [\forall z (z \notin y) \wedge y \in x \wedge \forall u R(u, x)]$ (where the relation R is defined above) is a sentence without parameters (it will later turn up in a slightly different form as the axiom of infinity).

The theorems of set theory (and the axioms in particular) are all closed formulas without parameters.

Equivalence Relations: A binary relation $R(x, y)$ is an *equivalence relation* if, whatever objects a, b, c might be,

$$R(a, b) \to R(a, a) \wedge R(b, b);$$
$$R(a, b) \to R(b, a);$$
$$R(a, b) \wedge R(b, c) \to R(a, c).$$

The class $R(x, x)$ is called the *domain* of the equivalence relation. $R(a, b)$ is also written $a \sim b$ (mod. R). For any set a in $\mathscr{U}$, the class $R(a, y)$ is called the *equivalence class* of a.

Ordering Relations (in the weak sense): A binary relation $R(x, y)$ is an *ordering relation* (or, briefly, an *ordering*) *in the weak sense* if for every a, b, and c

$$R(a, b) \to R(a, a) \wedge R(b, b);$$
$$R(a, b) \wedge R(b, a) \to a = b;$$
$$R(a, b) \wedge R(b, c) \to R(a, c).$$

The class $R(x, x)$ is the *domain* of the ordering. $R(a, b)$ is also written $a \leqslant b$ (mod. R); $R(a, b) \wedge b \neq a$ may similarly be abbreviated as $a < b$ (mod. R).

R is a (weak) *linear* (or *simple*, or *total*) ordering if, over and above all this, one or other of $R(a, b)$ and $R(b, a)$ holds for any two objects a, b in R's domain.

Ordering Relations (in the strict sense): A binary relation $R(x, y)$ defines a *strict ordering* on a class $D(x)$ if for all a, b, c,

$$R(a, b) \to D(a) \wedge D(b);$$
$$\neg(R(a, b) \wedge R(b, a));$$
$$R(a, b) \wedge R(b, c) \to R(a, c).$$

$R(a, b)$ may again be written $a < b$ (mod. R). It is obvious that if $R(x, y)$ is a strict ordering, then the relation $R(x, y) \vee [D(x) \wedge D(y) \wedge x = y]$ is a weak one, with domain D.

A strict ordering R on D is *linear* if for any objects a and b in the class D, either $R(a, b)$ or $R(b, a)$ or $a = b$.

Functional Relations: Consider by way of illustration a ternary relation $R(x, y, z)$. We will call it a *functional relation* of *two* arguments if

$$\forall x \forall y \forall z \forall z' [R(x, y, z) \wedge R(x, y, z') \to z = z'].$$

The binary relation $\exists z R(x, y, z)$ is then called the *domain* of the functional relation R. The class $\exists x \exists y R(x, y, z)$ is called the *range* of the functional relation R.

For any (for example binary) functional relation defined everywhere (in other words, a functional relation with domain $x=x \wedge y=y$) we can introduce a new symbol Φ, and write the functional relation as $z=\Phi(x, y)$. Then, for any formula $E(u, v, w)$, the formulas

$$\exists z [R(x, y, z) \wedge E(z, v, w)]$$

and

$$\forall z [R(x, y, z) \rightarrow E(z, v, w)]$$

(which are equivalent to one another) can be written simply as

$$E[\Phi(x, y), v, w].$$

For example, the functional relation

$$\forall t [t \in z \leftrightarrow t = x \vee t = y]$$

is written $z=\{x, y\}$; thus the equivalent formulas

$$\exists z [\forall t (t \in z \leftrightarrow t = x \vee t = y) \wedge z \in a]$$
$$\forall z [\forall t (t \in z \leftrightarrow t = x \vee t = y) \rightarrow z \in a]$$

can both be abbreviated by $\{x, y\} \in a$.

We can now go on to state the other axioms of set theory.

4. AXIOM SCHEME OF REPLACEMENT (OR SUBSTITUTION)

Suppose a formula $E(x, y, a_0, \ldots, a_{k-1})$ with parameters $a_0, \ldots, a_{k-1}$ defines a singulary functional relation; and let a be any set. Then we shall postulate that the universe $\mathscr{U}$ contains a set b whose elements are just the images under this functional relation of those elements of a within its domain. The demand on $\mathscr{U}$ that it satisfy this condition for any singulary functional relation is known as the axiom scheme of replacement. The scheme is rendered symbolically by the following infinite list of sentences:

$$\begin{aligned}\forall x_0 \ldots \forall x_{k-1} \{ \forall x \forall y \forall y' [E(x, y, x_0, \ldots, x_{k-1}) \\ \wedge E(x, y', x_0, \ldots, x_{k-1}) \rightarrow y = y'] \\ \rightarrow \forall t \exists w \forall v [v \in w \leftrightarrow \exists u [u \in t \wedge E(u, v, x_0, \ldots, x_{k-1})]]\}.\end{aligned}$$

Here $E(x, y, x_0, \ldots, x_{k-1})$ may be any parameter-free formula with at least two free variables x and y.

The axioms 1, 2, 3, the scheme 4, and the axiom of infinity (introduced on p. 29 below) make up the set theory of Zermelo and Fraenkel, usually abbreviated *ZF*.

Easily derived from the scheme of replacement is the following scheme.

Scheme of Comprehension: If a is a set, and $A(x, a_0, \ldots, a_{k-1})$ any formula of one free variable (and parameters $a_0, \ldots, a_{k-1}$), we can establish that there is a set whose elements are exactly those elements of a for which A holds. This is the content of the comprehension scheme, which consists of the following infinite list of sentences:

$$\forall x_0 \ldots \forall x_{k-1} \forall x \exists y \forall z \left[z \in y \leftrightarrow (z \in x \wedge A(z, x_0, \ldots, x_{k-1}))\right].$$

Here $A(x, x_0, \ldots, x_{k-1})$ is any parameter-free formula of at least one free variable x. To derive any instance of the scheme from the scheme of replacement it is enough to note that the formula $y = x \wedge A(x, a_0, \ldots, a_{k-1})$ defines a singulary functional relation F whose domain is the class $A(x, a_0, \ldots, a_{k-1})$. By replacement then, there is a set b whose elements are just the images under F of those members of a in F's domain; it is immediate that b is the set whose existence we set out to prove.

The set b is usefully represented by the notation

$$\{x \in a \mid A(x, a_0, \ldots, a_{k-1})\}.$$

THEOREM: *There is one and only one set without elements.*

PROOF: Let a be any set at all. We apply the comprehension scheme to the set a and the formula $x \neq x$ to get a set $b = \{x \in a \mid x \neq x\}$, which is obviously without elements. That there cannot be more than one such set follows at once from the axiom of extensionality. ∎

The set just defined is called the *empty set* and is denoted by $\emptyset$.

Let us give a derivation of the pairing axiom from axiom 3 and the scheme 4. First of all, every subset of $\emptyset$ is empty, so that $\mathscr{P}(\emptyset)$ has $\emptyset$ as its only element; thus $\{\emptyset\}$ exists. This latter set is a singleton, so if $a \subset \{\emptyset\}$ then $a = \emptyset$ or $a = \{\emptyset\}$. So $\mathscr{P}(\{\emptyset\})$ has just two distinct elements, $\emptyset$ and $\{\emptyset\}$; thus $\{\emptyset, \{\emptyset\}\}$ exists. Now let a and b be any sets you like. It is apparent that the relation $(x = \emptyset \wedge y = a) \vee (x = \{\emptyset\} \wedge y = b)$ is a singulary functional relation, so we can apply the replacement scheme to it and consider the image of the set $\{\emptyset, \{\emptyset\}\}$ thereunder. This turns out to be the set $\{a, b\}$, and thus the pairing axiom is proved.

We say that a class $A(x)$ *corresponds* to a set (or even, twisting language a bit, *is* a set) if there is a set a such that $\forall x(x \in a \leftrightarrow A(x))$. More generally, a relation (of three arguments, for example) $A(x, y, z)$ *corresponds* to a set, or even *is* a set, if there is a set a for which we have $\forall x\forall y\forall z[\langle x, y, z\rangle \in \in a \leftrightarrow A(x, y, z)]$.

It is not true that every class corresponds to a set; for example, the class $x \notin x$ does not. If it did, indeed, we would for some a have that $\forall x(x \in a \leftrightarrow x \notin x)$; so, in particular, $a \in a \leftrightarrow a \notin a$, which is evidently false. (This is Russell's paradox.)

The class $x = x$ (the universe $\mathscr{U}$, in other words) also fails to correspond to a set. For were there any set a such that $\forall x(x \in a)$ were true, we could, by using the comprehension scheme, produce a set b such that $\forall x(x \in b \leftrightarrow \leftrightarrow x \in a \wedge x \notin x)$; that is, $\forall x(x \in b \leftrightarrow x \notin x)$. So Russell's paradox would turn up all over again.

Cartesian Product of two sets: Consider the formula $X(z)$ abbreviated by

$$\exists x\exists y[z = \langle x, y\rangle \wedge x \in a \wedge y \in b].$$

To write this out in full we would first have to use the definition of the functional relation $z = \langle x, y\rangle$, obtaining

$$\exists x\exists y\exists u\exists v[z = \{u, v\} \wedge u = \{x\} \wedge v = \{x, y\} \\ \wedge x \in a \wedge y \in b].$$

The braces could then be eliminated with the definition of the functional relation $z = \{u, v\}$, leaving

$$\begin{aligned} \exists x\exists y\exists u\exists v[&\forall t(t \in z \leftrightarrow t = u \vee t = v) \\ &\wedge \forall t'(t' \in u \leftrightarrow t' = x) \\ &\wedge \forall t''(t'' \in v \leftrightarrow t'' = x \vee t'' = y) \\ &\wedge x \in a \wedge y \in b]. \end{aligned}$$

So $X(z)$ is just this last formula, and it clearly is the class of all ordered pairs $\langle x, y\rangle$ such that $x \in a$ and $y \in b$. (We will not bother to carry out this sort of check in the future; but it is suggested that the reader try a few, to get himself used to the business of manipulating formulas.)

Now this class X is in fact a set; for if $x \in a$ and $y \in b$ then (by the definition of $\langle x, y\rangle$) $\langle x, y\rangle \in \mathscr{P}(\mathscr{P}(a \cup b))$. But then $X(z)$ is equivalent to $X(z) \wedge z \in \mathscr{P}(\mathscr{P}(a \cup b))$, so that by the comprehension scheme X is a set.

This set X is called the *Cartesian* (or *direct*, or *cross*) *product* of the sets a, b, and it is written $a \times b$. The *Cartesian powers* $a \times a$, $a \times (a \times a)$, and so on, are sometimes abbreviated $a^2, a^3, \ldots$.

A singulary functional relation $R(x, y)$ whose domain is a set is itself a set. For suppose that the domain is a; then by the replacement scheme there is a set b consisting of the images of a under R. Consequently $R(x, y)$ is equivalent to $R(x, y) \wedge \langle x, y\rangle \in a \times b$, and this class is a set f by the comprehension scheme. Such a set f is called a *function defined on a, with values in b*, or a *map from a to b*, or a *family of sets indexed by a*. If a is a map we will sometimes write $dom(a)$, $rng(a)$ for its domain and range respectively.

The formula 'f is a map from a into b', written more compactly $f: a \to b$ is therefore just the following formula:

$$\begin{aligned} f \subset a \times b \quad & \wedge \forall x \forall y \forall y' [\langle x, y\rangle \in f \wedge \langle x, y'\rangle \in f \to y = y'] \\ & \wedge \forall x [x \in a \to \exists y (y \in b \wedge \langle x, y\rangle \in f)]. \end{aligned}$$

We leave to the reader the task of eliminating the defined symbols $\subset$, $\times$, and $\langle\ \rangle$.

If a and b are sets the class X of all maps from a into b is again a set. For a map from a to b is a subset of $a \times b$, and therefore a member of $\mathscr{P}(a \times b)$. So $X(f)$ is equivalent to $X(f) \wedge f \in \mathscr{P}(a \times b)$. Comprehension yields the desired result.

This set of maps is written ${}^a b$.

Union, Intersection, and Cartesian Product of a family of sets: Let a be a family of sets indexed by I; a is of course just a map whose domain is I, and to indicate this we will usually write a family as $(a_i)_{i \in I}$.

The *union of the family* $(a_i)_{i \in I}$, written $\bigcup_{i \in I} a_i$, is the union of the range of the map a. By the replacement scheme and union axiom it is a set b, and we have $\forall x [x \in b \leftrightarrow \exists i (i \in I \wedge x \in a_i)]$.

Likewise the *intersection of the family* $(a_i)_{i \in I}$ is the class $X(x)$ defined by $\forall i (i \in I \to x \in a_i)$. If $I = \emptyset$, then X is the class of all sets, and so not a set itself. Otherwise, take any $i_0 \in I$; then $X(x)$ is equivalent to $x \in a_{i_0} \wedge$

$\wedge\ \forall i(i \in I \to x \in a_i)$, and X is seen to be a set, by comprehension. We write it $\bigcap_{i \in I} a_i$.

Now take the class X of maps $f\colon I \to \bigcup_{i \in I} a_i$ such that $f(i) \in a_i$ for every $i \in I$. Such a map belongs to the set ${}^I(\bigcup_{i \in I} a_i)$, so $X(f)$ is equivalent to $X(f) \wedge f \in {}^I(\bigcup_{i \in I} a_i)$. X is thus a set, called the *Cartesian* (or *direct*, or *cross*) *product of the family* $(a_i)_{i \in I}$, and written $\bigtimes_{i \in I} a_i$ (or sometimes $\prod_{i \in I} a_i$).

Note that for the rest of the book we shall be using the words 'map' and 'function' in the sense defined above, which is by no means their every-day mathematical sense (since here, for example, all maps $f\colon a \to b$ are in the universe). As we proliferate definitions the same thing will happen to nearly all the familiar words of mathematical language. If we ever use the words in their normal senses (and this will not happen often), we will, as we decided before, say so *explicitly*. For example: let a and b be two sets, and A, B be the corresponding parts of the universe. If f maps a into b there will correspondingly be a map (in the intuitive sense) from A to B. But there may well be, intuitively speaking, maps from A to B to which there are no corresponding maps from a into b.

CHAPTER II

ORDINALS, CARDINALS

Well-ordering Relations: Let a be a set, all of whose elements lie in the domain of some linear ordering $R(x, y)$. We say that a is *well ordered* by R if every non-empty subset of a has a smallest element (mod. R). It is immediate that if a is well ordered by R then every subset of a is also well ordered by R.

If $r \subset a \times a$ then the pair $\langle a, r\rangle$ is said to be a *well-ordered system* if a is well ordered by the relation $\langle x, y\rangle \in r$.

Suppose that the set a is well ordered by the relation R. A subset s of a is said to be an *initial segment* of a if $x \in s$ and $y \leqslant x$ (mod. R) imply $y \in s$, for any $x, y \in a$. For each x_0 in a we write $S_{x_0}(a, R)$ (or $S_{x_0}(a)$ if there is no danger of ambiguity) for the set $\{x \in a \mid x < x_0 \text{ (mod. } R)\}$, and this set is evidently an initial segment of a.

Indeed, *a subset $s \subset a$ is an initial segment of a iff $s = a$ or $s = S_{x_0}(a)$ for some $x_0 \in a$.*

This is easily proved by noting that if s is an initial segment of a, but not a itself, then the set $a \setminus s = \{x \in a \mid x \notin s\}$ is non-empty. Consequently, it has a least element x_0; so if $x < x_0$, x must be in s; and conversely, if $x \geqslant x_0$, we cannot have $x \in s$ (for that implies that $x_0 \in s$).

An initial segment of a which is not just a itself is called a *proper initial segment* of a.

If R is a linear ordering, and x is in the domain of R, let $S_x(R)$ stand for the class of y's in the domain of R which are $< x$ (mod. R). Then R is called a *well-ordering* (relation) if, for any x in its domain, the class $S_x(R)$ is a set, and is also well ordered by R.

If R is a well-ordering with domain D, and T any non-empty subclass of D (that is, $\forall x(T(x) \to D(x)) \wedge \exists x T(x)$), then T has a least element (mod. R).

For T is non-empty, so take x_0 in T. Either x_0 is the least element of T, or else the class $T \wedge S_{x_0}(R)$ is non-empty. This class is however a set (since $S_{x_0}(R)$ is a set), and indeed a non-empty subset of $S_{x_0}(R)$, which is well ordered. It therefore has a least element, which is obviously also the least element of T.

(Our talk here of elements of the class T is to be understood as using 'element' in an informal way only.)

The Class of Ordinals: A set α is called *transitive* if every element of α is also a subset of α; that is, if $\forall x(x \in \alpha \rightarrow x \subset \alpha)$. A set α is called an *ordinal* if, in addition to being transitive, it is strictly well ordered by the membership relation $x \in y$.

The class 'α is an ordinal' is written $On(\alpha)$; spelt out more or less in full this becomes

$$\begin{aligned}&\forall x\forall y[x \in \alpha \wedge y \in \alpha \rightarrow x \notin y \vee y \notin x]\\ &\wedge \forall x\forall y\forall z[x \in \alpha \wedge y \in \alpha \wedge z \in \alpha \wedge x \in y \wedge y \in z \rightarrow x \in z]\\ &\wedge \forall z[z \subset \alpha \wedge z \neq \emptyset \rightarrow \exists x(x \in z\\ &\qquad\qquad\qquad\qquad \wedge \forall y(y \in z \rightarrow x \in y \vee x = y))]\\ &\wedge \forall x\forall y[x \in \alpha \wedge y \in x \rightarrow y \in \alpha].\end{aligned}$$

(In this extensive formula there is no explicit mention of the fact that the ordering $x \in y$ is linear; however, no such clause is needed, since we have stated that every non-empty subset of α contains a smallest element, and *a fortiori* this applies to subsets of two elements.)

Examples of ordinals that are easily shown to be such are $\emptyset$, $\{\emptyset\}$, $\{\emptyset, \{\emptyset\}\}$.

Let α be an ordinal. Then the initial segments of α are α itself, and the elements of α.

For a proper initial segment of α is $S_\xi(\alpha)$ for some $\xi \in \alpha$; and $S_\xi(\alpha) = \{\eta \in \alpha \mid \eta < \xi\} = \{\eta \in \alpha \mid \eta \in \xi\} = \xi \cap \alpha$. But since $\xi \subset \alpha$, this last term reduces to ξ, and the result is proved.

Every element of an ordinal is an ordinal.

Take $\xi \in \alpha$; then $\xi \subset \alpha$, so membership well-orders ξ. Furthermore, if $y \in x$ and $x \in \xi$ then $x \in \alpha$ (since $\xi \subset \alpha$); so $x \subset \alpha$, so $y \in \alpha$. Now membership strictly orders the elements of α, so from $y \in x$ and $x \in \xi$ we get $y \in \xi$. Thus $x \in \xi$ implies that $x \subset \xi$, and the two conditions for ξ to be an ordinal are established.

For every ordinal α, $\alpha \notin \alpha$.

If ξ is any element of α, then $\xi \notin \xi$, since $\in$ strictly orders α. So if α is an element of α we must have $\alpha \notin \alpha$. Thus $\alpha \notin \alpha$.

For every two ordinals α, β, either $\alpha = \beta$, or $\beta \in \alpha$, or $\alpha \in \beta$; moreover, the three cases are mutually exclusive.

Put $\xi = \alpha \cap \beta$. Then ξ consists of all members of α that are $<\beta$ (mod. $\in$), and so is an initial segment of α. Thus $\xi = \alpha$ or $\xi \in \alpha$. Likewise ξ is an initial segment of β, so $\xi = \beta$ or $\xi \in \beta$. All in all then, there are four possibilities:

$\xi = \alpha$ and $\xi = \beta$; then $\alpha = \beta$.
$\xi = \alpha$ and $\xi \in \beta$; then $\alpha \in \beta$.
$\xi = \beta$ and $\xi \in \alpha$; then $\beta \in \alpha$.
$\xi \in \alpha$ and $\xi \in \beta$; then $\xi \in \alpha \cap \beta$, so by the definition of ξ, $\xi \in \xi$. But this is a contradiction, since ξ is an ordinal.

Furthermore, $\alpha = \beta$ and $\alpha \in \beta$ together would imply that $\beta \in \beta$; similarly, if $\alpha \in \beta$, then $\alpha \subset \beta$, so if $\beta \in \alpha$ were also true, we would have $\beta \in \beta$ again. But $\beta \notin \beta$. Thus the three cases are mutually exclusive.

The membership relation on the class On (that is, the relation $x \in y \wedge \wedge On(x) \wedge On(y)$*) is a well-ordering.*

We have just shown the relation in question to be a strict linear ordering. So we need only note that since $\xi < \alpha \leftrightarrow \xi \in \alpha$, $S_\alpha(\in) = \alpha$ for every ordinal α, and α is well ordered by $\in$.

Observe that if α, β *are ordinals then* $\alpha \leqslant \beta \leftrightarrow \alpha \subset \beta$.

On is not a set.

Suppose *On* is a set a. Then a is well ordered by $\in$. Moreover, every element of a is an ordinal, therefore a set of ordinals, therefore a subset of a. But then a qualifies as an ordinal; so $a \in a$. But this is impossible for ordinals.

If α *is an ordinal the least ordinal greater than* α *is* $\alpha \cup \{\alpha\}$. *This is called the successor of* α, *and written* $\alpha + 1$.

It is easily checked that if α is an ordinal then so is $\beta = \alpha \cup \{\alpha\}$. Now since $\alpha \in \beta$, we have $\alpha < \beta$; and if $\alpha < \gamma$ then $\alpha \in \gamma$, so $\alpha \subset \gamma$, so $\alpha \cup \{\alpha\} \subset \gamma$. Thus $\beta \leqslant \gamma$.

Since $\emptyset \subset \alpha$ for any α, the ordinal $\emptyset$ is the least ordinal of all. We call it 0. Its successor $\{\emptyset\}$ is called 1; *its* successor $1 \cup \{1\}$ (that is, $\{\emptyset, \{\emptyset\}\}$) is called 2; and so on.

The union of any set of ordinals is the least upper bound of the set.

Let a be any set of ordinals, and $\beta = \bigcup_{\alpha \in a} \alpha$. If x is a non-empty subset of β, then for at least one $\alpha_0 \in a$ we will have $x \cap \alpha_0$ non-empty. Since α_0 is well ordered, its subset $x \cap \alpha_0$ will therefore have a least element, which is clearly also the least element of x. Since x was an arbitrary non-empty subset of β, it follows that β is well ordered by $\in$. Now suppose that $y \in x$ and $x \in \beta$. Then $x \in \alpha$ for some $\alpha \in a$, so $y \in \alpha$ since α is an ordinal. We conclude that $y \in \beta$, and this establishes that β is transitive. It is thus an ordinal. Moreover $\alpha \leqslant \beta$ for any $\alpha \in a$, so β is an upper bound of a. And if $\alpha \leqslant \gamma$ for any $\alpha \in a$, then $\alpha \subset \gamma$ for any $\alpha \in a$, so $\beta \subset \gamma$, so $\beta \leqslant \gamma$. Thus β is the least upper bound of a.

THEOREM: *Let α, β be ordinals, and f an order-preserving isomorphism from α on to β. Then $\alpha = \beta$, and f is the identity map.*
PROOF: Suppose f is not the identity, and let ξ be the least member of α for which $f(\xi) \neq \xi$. Then $\eta \in \xi \to f(\eta) = \eta$, so $\xi \subset \beta$. Being an ordinal, ξ is an initial segment of β; but it is not β itself, for in that case f would map a proper initial segment of α on to the whole of β, which is not possible if f is an isomorphism. Thus $\xi \in \beta$.

Since f is order-preserving, $\eta < \xi \to f(\eta) < f(\xi)$; by the definition of ξ therefore, $\eta < f(\xi)$ for every $\eta \in \xi$. Thus $\xi \subset f(\xi)$, or, equivalently, $\xi \leqslant f(\xi)$. Equality is ruled out, so $\xi < f(\xi)$. We now ask which element of α is mapped by f to ξ (which is in β, as proved above). It is easily seen that there cannot be one, since $x \geqslant \xi \to f(x) > \xi$ and $x < \xi \to f(x) < \xi$; thus f is not an isomorphism from α on to β. ∎

THEOREM: *For each well-ordered system $u = \langle a, r \rangle$ there exists exactly one (order-preserving) isomorphism from a on to an ordinal.*
PROOF: (*Uniqueness*) Suppose f maps a one-one on to α, and g does the same on to β. Then $g \circ f^{-1}$ is an isomorphism from α on to β, and it can only be the identity. So $f = g$.

(*Existence*) Put $b = \{x \in a \mid S_x(a) \text{ is isomorphic to an ordinal}\}$. By uniqueness no such $S_x(a)$ can be isomorphic to more than one ordinal, so let $\beta(x)$ be the ordinal that $S_x(a)$ is isomorphic to.

If $y \in b$ and $x < y$ then x must be in b, and, indeed, $\beta(x) < \beta(y)$; for under the isomorphism from $S_y(a)$ on to $\beta(y)$, the proper initial segment $S_x(a)$ of $S_y(a)$ becomes a proper initial segment of $\beta(y)$, and therefore an ordinal less than $\beta(y)$, therefore $\beta(x)$. Thus b is an initial segment of a.

By replacement we can form the set β of those ordinals $\beta(x)$ where $x \in b$. Now $\xi \leqslant \beta(x)$ implies that ξ is isomorphic to an initial segment of $S_x(a)$, say $S_y(a)$ for some $y \leqslant x$ (so that $\xi = \beta(y)$). It follows that β is an initial segment of the ordinals, and therefore itself an ordinal.

Thus the map sending x to $\beta(x)$ is an isomorphism from b on to β.

Our existence proof will be complete if we can show that $b=a$. And this is simply done. For suppose $b \neq a$; since b is an initial segment of a, it must be true for some $x_0 \in a$ that $b=S_{x_0}(a)$. We have just proved that b is isomorphic to an ordinal, so it follows from the definition of b that $x_0 \in b$; but it is impossible that $x_0 \in S_{x_0}(a)$, by the very definition of initial segment. Thus $b=a$ after all. ∎

Consider now any well-ordering $R(x, y)$ whose domain D is not a set; we shall define a functional relation J which establishes an isomorphism between D and On.

The functional relation $J(x)=\alpha$ is defined by the formula 'α is an ordinal isomorphic to the segment $S_x(R)$'. This is a functional relation all right, and its domain is D; for whatever x we pick in D there is one and only one ordinal isomorphic to $S_x(R)$. It is also an order-preserving isomorphism. For if $x<y$ (mod. R) then $S_x(R)$ is a proper initial segment of $S_y(R)$. But $S_y(R)$ can be isomorphically mapped on to $J(y)$, and the image of $S_x(R)$ under this mapping will form a proper initial segment of $J(y)$, which is thus an ordinal $J(x)<J(y)$.

Furthermore, the range of J is an initial segment of On; for if $\beta \leqslant J(x)$ then when $J(x)$ is mapped on to $S_x(R)$, β is mapped on to an initial segment of the latter, $S_y(R)$ say, for some $y \leqslant x$. Thus $\beta=J(y)$. Since J is one-one it has an inverse J^{-1} which maps the range of J into D. But D is not a set, so by replacement the domain of J^{-1} is not a set. Thus the range of J is some initial segment of On; but not a set.

Under these circumstances it can only be On itself, which establishes that J is a surjection.

Note that no other functional relation J' could be used to establish an isomorphism between D and On. For whatever x in D we choose, $J'(x)$ is an ordinal isomorphic to $S_x(R)$; only one ordinal has this property, and this is $J(x)$. Thus $J'(x)=J(x)$ at every point x in D.

Proof by Induction on the Ordinals: Let $E(x)$ be some formula containing only the variable x free. If $E(\alpha)$ is to hold for every ordinal α, it is necessary and sufficient that $E(\alpha)$ holds whenever $E(\beta)$ holds for every $\beta<\alpha$; more succinctly, that

$$\forall\alpha[\forall\beta(\beta<\alpha \rightarrow E(\beta)) \rightarrow E(\alpha)]$$

holds generally.

The proof falls out at once. If $E(\alpha)$ is false for any ordinal at all, we let α be the least such ordinal. Then $E(\beta)$ is true whenever $\beta<\alpha$, and so by hypothesis $E(\alpha)$ is true, which is a contradiction.

This method of proof of $\forall\alpha E(\alpha)$ is called *proof by induction.*

Definition by Recursion on the Ordinals: Suppose that F is a functional relation and T a subclass of F's domain. That functional relation $y= =F(x) \wedge T(x)$ which restricts F to T is written $F \upharpoonright T$; in particular, if a is a subset of the domain of F, then $F \upharpoonright a$ represents a map, the *restriction* of F to a.

THEOREM: *Let α be an ordinal; A a class; M the class of all maps defined on ordinals less than α, and taking values in A; and H a functional relation with domain M, and taking values in A. Then there exists one and only one map f defined on α for which $f(\beta)=H(f \upharpoonright \beta)$ for every $\beta \in \alpha$.*

PROOF: (*Uniqueness*) This is routine. Suppose distinct functions f and g both did the job required, and let β be the least ordinal in α where $f(\beta)\neq g(\beta)$. Then $f(\gamma)=g(\gamma)$ for $\gamma<\beta$, so $f \upharpoonright \beta=g \upharpoonright \beta$. But this means that $f(\beta)=H(f \upharpoonright \beta)=H(g \upharpoonright \beta)=g(\beta)$, which is impossible.

(*Existence*) Let τ be the set of all $\beta<\alpha$ on which there is defined a map f_β satisfying $f_\beta(\gamma)=H(f_\beta \upharpoonright \gamma)$ for every $\gamma<\beta$.

It is clear that τ is an initial segment of α. By the uniqueness part of the theorem, indeed, we actually have $f_\beta=f_{\beta'} \upharpoonright \beta$ for $\beta<\beta' \in \tau$. Thus τ is an ordinal $\leqslant\alpha$, and we can define a map f_τ on τ as follows: $f_\tau(\beta)= =H(f_\beta)$, for every $\beta \in \tau$.

Suppose $\gamma \in \beta \in \tau$; then $f_\tau(\gamma)=H(f_\gamma)=H(f_\beta \upharpoonright \gamma)=f_\beta(\gamma)$. Thus $f_\beta=f_\tau \upharpoonright \beta$. Consequently, $f_\tau(\beta)=H(f_\tau \upharpoonright \beta)$ for all $\beta \in \tau$.

But this means that if $\tau<\alpha$ then it satisfies the defining property of τ; in other words that $\tau \in \tau$, which is impossible. Thus τ, which is an initial segment of α, is actually equal to α, and so f_τ is the function we are looking for. ∎

Let A be a class; M the class of all maps defined on ordinals, and taking values in A; and H a functional relation with domain M, and taking values in A. We can now define a unique functional relation F with domain On, and taking values in A, which for every ordinal α satisfies $F(\alpha)=H(F \upharpoonright \alpha)$.

The functional relation $y=F(\alpha)$ is supplied by the formula 'there is a map f_α defined on α such that, for all $\beta<\alpha$, $f_\alpha(\beta)=H(f_\alpha \restriction \beta)$, and $y=H(f_\alpha)$'.

We have just proved that for each ordinal α there is a unique map f_α of this sort, and this establishes that the relation just defined is a functional relation with domain *On*.

To prove that F has the right values, take $\beta<\alpha$. Then $f_\beta=f_\alpha \restriction \beta$ simply because the latter map has all the properties defining f_β. Thus

$$F(\beta)=H(f_\beta)=H(f_\alpha \restriction \beta)=f_\alpha(\beta)$$

for all $\beta<\alpha$. Manifestly then, $f_\alpha=F\restriction\alpha$; but $F(\alpha)=H(f_\alpha)$, so $F(\alpha)= =H(F\restriction\alpha)$.

That F is the only such functional relation is proved in the same way as before. Suppose G is another, and let α be the least ordinal where F and G differ (if there is one). Since $F(\beta)=G(\beta)$ for all $\beta<\alpha$ we must have $F\restriction\alpha=G\restriction\alpha$, so $H(F\restriction\alpha)=H(G\restriction\alpha)$. Thus $F(\alpha)=G(\alpha)$, a contradiction. It follows that F and G are equal for every ordinal α.

THE AXIOM OF CHOICE

The axiom of choice (*AC* for short) is the following assertion:

If a is a set of pairwise disjoint but non-empty sets, then there is a set whose intersections with elements of a are always singletons.

Spelt out more fully this becomes

$$\begin{aligned}\forall a\,\{[\forall x(x\in a\rightarrow x\neq\emptyset)\\ &\wedge \forall x\forall y(x\in a\wedge y\in a\rightarrow(x=y\vee x\cap y=\emptyset))]\\ &\rightarrow \exists b\forall x\exists u(x\in a\rightarrow b\cap x=\{u\})\}\,.\end{aligned}$$

Given the other axioms the following statements are equivalent to the axiom of choice.

AC': *For any set a there is a function h mapping the set of all non-empty subsets of a into a in such a way that $h(x)\in x$ for all non-empty $x\subset a$.*

AC'': *The Cartesian product of a family of non-empty sets is itself non-empty.*

$AC'\rightarrow AC$: Let $b=\bigcup a$. We apply AC' to b. Every element x of a, being non-empty, is a non-empty subset of b. So take the function h that AC' guarantees, and consider $\{h(x)\mid x\in a\}$. This set has exactly one element from each $x\in a$, and is the set needed to establish AC.

$AC \to AC''$: Let $(a_i)_{i \in I}$ be a family of non-empty sets, and put $b_i = \{i\} \times a_i$. Then the family $(b_i)_{i \in I}$ consists throughout of pairwise disjoint non-empty sets. Let ξ be a set with just one element in each $b_i (i \in I)$. Then $\xi \in \bigtimes_{i \in I} a_i$, by the definition of $\bigtimes$.

$AC'' \to AC'$: Take any set a and form the product $\bigtimes_{\substack{x \subset a \\ x \neq \emptyset}} x$. By AC'' this is not empty, so choose an element φ. It is easily checked that $\varphi(x) \in x$ for every non-empty subset x of a.

A further equivalent of the axiom of choice is the following statement.

ZERMELO'S THEOREM: *Every set can be well ordered.*

PROOF: AC' follows at once from this. For take some $r \subset a^2$ that well-orders a and define a map h: $\mathscr{P}(a) \setminus \{\emptyset\} \to a$ by putting $h(x) =$ the least element of x (mod. r).

Conversely, suppose that AC' is true and that the map

$$h: \mathscr{P}(a) \setminus \{\emptyset\} \to a$$

satisfies the requirement that $h(x) \in x$ for every non-empty $x \subset a$. Define a second map

$$g: \mathscr{P}(a) \setminus \{a\} \to a$$

by putting $g(x) = h(a \setminus x)$. Then this satisfies the requirement that $g(x) \notin x$ for any proper $x \subset a$.

Let θ be some object that is not an element of a (since the class of all sets is not itself a set, there is no difficulty about finding such a θ). We define by recursion a functional relation F: $On \to a \cup \{\theta\}$ by decreeing

$$F(\alpha) = \begin{cases} g(\{F(\beta) \mid \beta < \alpha\}) \quad \text{if} \quad \{F(\beta) \mid \beta < \alpha\} \subsetneq a \\ \text{(that is, if it is in the domain of } g\text{);} \\ \theta \text{ otherwise}. \end{cases}$$

Suppose that F never takes the value θ; in other words, that $F(\alpha) \in a$ for every ordinal α. Then $\{F(\beta) \mid \beta < \alpha\}$ is in g's domain, whatever ordinal α is. But $F(\alpha)$ cannot equal any $F(\beta)$ for $\beta < \alpha$, because

$$F(\alpha) = g(\{F(\beta) \mid \beta < \alpha\}) \notin \{F(\beta) \mid \beta < \alpha\}$$

by the definition of g. Thus F is an injection; On is its domain, and all its values are in a. Inverting F, therefore, and using replacement, we can infer that *On* is a set. Since this is false we must conclude that $F(\alpha)$ is not always in a.

Thus $F(\alpha)=\theta$ for some ordinal α, and we may suppose α_0 to be the least such. In this case $F(\beta)\in a$ for every $\beta<\alpha_0$, and so $\{F(\beta)\mid\beta<\alpha_0\}$ is a subset of a; yet it is not in the domain of g (since $F(\alpha_0)=\theta$). By the definition of g, $\{F(\beta)\mid\beta<\alpha_0\}$ can then only be a itself; so the range of $F\restriction\alpha_0$ is a.

But $F\restriction\alpha_0$ is also an injection; for if $\beta<\alpha_0$, $F(\beta)=g(\{F(\gamma)\mid\gamma<\beta\})\notin \notin\{F(\gamma)\mid\gamma<\beta\}$, so that (as before) $F(\beta)\neq F(\gamma)$ when $\gamma<\beta$.

Thus $F\restriction\alpha_0$ is a bijection from the ordinal α_0 on to a, and consequently a can be well ordered. ∎

Note that what we have just proved can be stated as follows: whether or not *AC* holds, *a set can be well ordered if and only if there is a map* h: $\mathscr{P}(a)\diagdown\{\emptyset\}\to a$ *for which* $h(x)\in x$ *for every non-empty* $x\subset a$.

DEFINITION: If R is an ordering and T a subclass of the domain D of R, then an object x of D is called an *upper bound* (respectively, a *strict upper bound*) of T if, for every $y\in T$, we have $x\geqslant y$ (respectively, $x>y$). If T has a strict upper bound we call it *dominated.* If the class of upper bounds of T has a smallest element under R, then this element is called the *least upper bound* of T.

An object x in D is called a *maximal element* of D if there exists no greater element of D (under the ordering R).

We now prove yet another equivalent of the axiom of choice.

ZORN'S LEMMA: *Let u be an ordered system, every well-ordered subset of which has an upper bound. Then u has a maximal element.*

PROOF: Put $u=\langle a, r\rangle$ where $r\subset a^2$ is an ordering. Assuming the axiom of choice in the form AC' we have a map h: $\mathscr{P}(a)\diagdown\{\emptyset\}\to a$ for which $h(x)\in x$ for every non-empty $x\subset a$.

Let c be the set of all dominated subsets of a (subsets with *strict* upper bounds). We define a map m from c into a by putting

$$m(x) = h\,(\text{the set of strict upper bounds of } x)\,.$$

Then for $x\in c$, $m(x)$ itself is a strict upper bound of x, and so $m(x)$ cannot be a member of x.

Choose again some θ which is not a member of a. As in the proof of

Zermelo's theorem, we can define a functional relation $F: On \to a \cup \{\theta\}$ as follows:

$$F(\alpha) = \begin{cases} m\{F(\beta) \mid \beta < \alpha\} & \text{if } \{F(\beta) \mid \beta < \alpha\} \in c; \\ \theta \text{ otherwise (that is, if } \{F(\beta) \mid \beta < \alpha\} \\ \text{has no strict upper bound or is not included in } a). \end{cases}$$

Again we show that F must take the value θ at least once. For, were it otherwise, and $F(\alpha) \in a$ for every ordinal α, we would have $\{F(\beta) \mid \beta < \alpha\} \in c$ for every ordinal α. Moreover, since $F(\alpha)$ dominates $\{F(\beta) \mid \beta < \alpha\}$, it cannot belong to this set, and so for $\beta < \alpha$, we get $F(\beta) \neq F(\alpha)$. Thus F is one-one; once more the scheme of replacement applied to F^{-1} would deliver On as a set, and so we have a contradiction.

So let α_0 be the least ordinal for which $F(\alpha_0) = \theta$. Any smaller ordinal α will satisfy $F(\alpha) \in a$; and so $\{F(\beta) \mid \beta < \alpha\} \in c$; thus $F(\alpha)$ will be a strict upper bound to $\{F(\beta) \mid \beta < \alpha\}$, and $\beta < \alpha < \alpha_0 \to F(\beta) < F(\alpha)$ (mod. r). It follows that $F \restriction \alpha_0$, which maps α_0 one-one into a, is also order-preserving. Consequently $\{F(\beta) \mid \beta < \alpha_0\}$ is going to be a well-ordered subset of a, and so by hypothesis it has an upper bound, d say. But it has no *strict* upper bound, for the existence of such an object would put $\{F(\beta) \mid \beta < \alpha_0\}$ in c; and this would mean that $F(\alpha_0) \in a$, contrary to supposition.

Thus there is no element of a that is greater than d (under the ordering r); and so d is the desired maximal element of a. ▌

Conversely, we can show that a seemingly weaker statement than Zorn's lemma is sufficient to entail the axiom of choice.

CONVERSE: *Suppose that there is a maximal element in every ordered system whose linearly ordered subsets all have least upper bounds. Then the axiom of choice is true.*

PROOF: Let a be a set of non-empty pairwise disjoint sets; and let $b = \bigcup a$. Put X equal to the set of all subsets of b which do not have more than one element in common with any $x \in a$; then X is ordered by set inclusion.

Suppose Y is any linearly ordered subset of X. We shall show that Y has a least upper bound in X. Now the least upper bound of Y (under the $\subset$-ordering) is $\bigcup_{y \in Y} y$: so we will show that $\bigcup_{y \in Y} y \in X$.

For any x in a, then, the set $x \cap \bigcup_{y \in Y} y$ must be proved to have fewer than two elements. There are two cases to be considered. In the first, $x \cap y = \emptyset$ for every $y \in Y$; here we have at once $x \cap \bigcup_{y \in Y} y = \emptyset$, and nothing remains to be proved. In the second, $x \cap y$ is a singleton for one or more $y \in Y$. But however many $y \in Y$ yield a singleton when intersected with x, *it will always be the same singleton.* For suppose $x \cap y_0 = \{u\}$ and $x \cap y_1 = \{v\}$. Then since Y is linearly ordered, $y_0 \cup y_1$ is one or the other of y_0, y_1, which means that $x \cap (y_0 \cup y_1) = \{u\} \cup \{v\}$ is a singleton; and this means that $u = v$. It follows at once from all this that, in this second case, $x \cap \bigcup_{y \in Y} y$ is itself a singleton, so that $\bigcup_{y \in Y} y \in X$, as was required.

By supposition, then, X has a maximal element y_0. We claim that $y_0 \cap x$ has exactly one element for each $x \in a$. Were it otherwise, indeed, $y_0 \cap x$ would have to have no elements for some $x \in a$ (since $y_0 \in X$). In this case we could take any element ξ of x and form the set $y_0 \cup \{\xi\}$. This, however, would equally be in X, and would destroy y_0's maximality. ∎

CARDINALS

(In this section the axiom of choice will be used.)

Two sets a and b are termed *equipollent* (or *equinumerous*, or sometimes just *equivalent*) if there is a one-one map (a bijection) from a on to b.

It is to be observed that two sets a, b can be equipollent in an intuitive sense without being so according to this definition; for if A, B are the parts of the universe corresponding to a, b, there may be a bijection, intuitively speaking, from A on to B. But unless this bijection itself corresponds to an object of the universe there is no reason to suppose that a and b are therefore equipollent.

It is obvious enough that the relation 'x is equipollent to y' is an equivalence relation whose domain is the class of all sets. With the axiom of choice, however, we can say more than this; namely that *every set is equipollent to an ordinal.* For if a set can be well ordered, it is certainly equipollent to the ordinal of its well-ordering.

The least ordinal equipollent to a set a, written $\bar{a}$, or sometimes *card*(a), is called the *cardinal* of the set a. It is trivial that a and b are equipollent if and only if $\bar{a} = \bar{b}$.

We write *Card* for the class of all cardinals. It is defined by the formula

Card(α): 'α is an ordinal not equipollent to any smaller ordinal'.

LEMMA: *Let α be an ordinal and ξ a subset of it. Then the ordinal of the well-ordering induced on ξ by the ordering of α is less than or equal to α.*
PROOF: Let β be the ordinal in question, and $f\colon \xi \to \beta$ the isomorphism between ξ and β. We show that for $\gamma \in \xi$, $f(\gamma) \leqslant \gamma$, which immediately gives $\beta \subset \alpha$, and so $\beta \leqslant \alpha$ as required.

Suppose, on the contrary, that $f(\gamma) > \gamma$ for some $\gamma \in \xi$. There will be a least such, γ_0. For $\gamma \in \xi$ we have then

$$\gamma < \gamma_0 \to f(\gamma) \leqslant \gamma < \gamma_0\,;$$

and also

$$\gamma \geqslant \gamma_0 \to f(\gamma) \geqslant f(\gamma_0) > \gamma_0\,.$$

As in an earlier theorem, f clearly misses out on γ_0; but since it does take greater ordinals as values (for example, $f(\gamma_0) > \gamma_0$), it cannot after all have an ordinal for its range; and this contradicts the hypothesis. ▮

THEOREM: *Let a and b be non-empty sets. Then the following conditions are equivalent.*

(1) *There is a one-one map from a into b.*
(2) *There is a map from b on to a.*
(3) $\bar{\bar{a}} \leqslant \bar{\bar{b}}$.

PROOF: ((1) → (2)) Let j be an injection from a into b. We define a surjection s in the opposite direction as follows. Take any x_0 in a, and let $s(y)$ be x_0 for any y in b that is not in the range of j; and for elements y of b that are in the range of j we can take $s(y)$ as the inverse image of y under j, since j is one-one on its range.

((2) → (1)) Conversely, suppose $s\colon b \to a$ is a surjection. *AC* gives us a map $h\colon \mathscr{P}(b) \setminus \{\emptyset\} \to b$ such that $h(Y) \in Y$ for every non-empty $Y \subset b$. So for $x \in a$ we put $j(x) = h(\{y \in b \mid s(y) = x\})$. This is manifestly one-one.

((3) → (1)) If $\bar{\bar{a}} \leqslant \bar{\bar{b}}$ we have two bijections $f\colon a \to \bar{\bar{a}}$ and $g\colon b \to \bar{\bar{b}}$; and, since $\bar{\bar{a}} \subset \bar{\bar{b}}$, an identity injection $i\colon \bar{\bar{a}} \to \bar{\bar{b}}$. Composing these into $g^{-1} \circ i \circ f$, we produce what is clearly an injection from a into b.

((1) → (3)) Suppose $j\colon a \to b$ is an injection. Then, with f and g defined as above, the map $g \circ j \circ f^{-1}$ is an injection k from $\bar{\bar{a}}$ into $\bar{\bar{b}}$. The range of k is some subset $\xi \subset \bar{\bar{b}}$ which, according to the previous lemma, is isomorphic to some ordinal $\beta \leqslant \bar{b}$. Thus $\bar{a}$ is equipollent with β, and since it is a cardinal, $\bar{a} \leqslant \beta$; so $\bar{a} \leqslant \bar{b}$. ▮

COROLLARY (CANTOR/BERNSTEIN THEOREM): *For two sets a, b to be equipollent it is necessary and sufficient that there be injections from each one into the other.*

PROOF: For given the two injections we get at once $\bar{a} \leqslant \bar{b}$ and $\bar{b} \leqslant \bar{a}$, and thus $\bar{a} = \bar{b}$. ▮

(Although the axiom of choice was used in this theorem, and thereby in the proof of the corollary, it could, in the latter case, have been dispensed with.)

THEOREM (CANTOR): *For every set a, $\bar{a} < \overline{\overline{\mathscr{P}(a)}}$.*

PROOF: Suppose that $\bar{a} \geqslant \overline{\overline{\mathscr{P}(a)}}$; then there exists a map h from a on to $\mathscr{P}(a)$. Let $b = \{x \in a \mid x \notin h(x)\}$. Then $b \subset a$, so there is a $c \in a$ such that $h(c) = b$. But $c \in b$ is equivalent to $c \notin h(c)$, and this is equivalent to $c \notin b$; and this is a contradiction. ▮

Note that in proving this we have also proved the following theorem. For neither it nor the next one do we need rely on AC.

THEOREM: *For no set a does there exist a map from a on to $\mathscr{P}(a)$; and,* a fortiori, *no injection of $\mathscr{P}(a)$ into a.*

THEOREM: *The class of cardinals is not a set.*

PROOF: Suppose it were, x for example. Then x would be a set of ordinals with a least upper bound $\lambda = \bigcup_{\alpha \in x} \alpha$. Since every ordinal is equipollent with some cardinal or another, every ordinal is equipollent with some subset of λ. Now let Y be the class of all well-orderings defined on subsets of λ. Every $r \in Y$ is a subset of λ^2; so $Y(r)$ is equivalent to the formula $r \in \mathscr{P}(\lambda^2) \wedge Y(r)$, which establishes that Y is a set (by the scheme of comprehension).

But every r in Y is associated with a unique ordinal (namely, *the* ordinal of the well-ordering r). Consequently, we can write down a functional relation performing this association, and the range of this functional relation is On. Since Y is a set, this contradicts the replacement scheme. ▮

Let a, b be two disjoint sets, and a', b' two more, respectively equipollent with a and b. Then $a \cup b$ and $a' \cup b'$ are equipollent too; for from the first to the second there is a bijection whose restrictions to a and b are just the given bijections from a to a' and from b to b'.

So let α, β be cardinals. There certainly exist disjoint sets a, b with these cardinalities, for we can, if we like, take $\alpha \times \{0\}$ for a and $\beta \times \{1\}$ for b. Whatever disjoint a and b we actually choose, we define the *cardinal sum* $\alpha + \beta$ as the cardinal of $a \cup b$; it is easily seen that $\alpha + \beta$ is quite independent of the choice of a and b.

Amongst the properties easily proved to hold for cardinal addition are commutativity $(\alpha + \beta = \beta + \alpha)$ and associativity $(\alpha + (\beta + \gamma) = (\alpha + \beta) + \gamma)$.

To get at cardinal multiplication we can drop the disjointness conditions on a, b and a', b' that were imposed above. If $f: a \to a'$ and $g: b \to b'$ are some appropriate bijections, then we can define a bijection $h: a \times b \to a' \times b'$ by writing $h(\langle x, y \rangle) = \langle f(x), g(y) \rangle$ for all $x \in a$ and $y \in b$; thus these Cartesian products are equipollent.

So if α, β are cardinals we define the *(cardinal) product* of α and β, written for brevity $\alpha\beta$, or $\alpha \cdot \beta$, as the cardinal of $\alpha \times \beta$. It should be clear that if a has cardinal α and b has cardinal β, then $a \times b$ has cardinal $\alpha\beta$.

If α, β, γ are all cardinals we can easily check that the commutative $(\alpha\beta = \beta\alpha)$ and associative $(\alpha(\beta\gamma) = (\alpha\beta)\gamma)$ laws of multiplication hold; there is no more difficulty involved in checking the distributive law $(\alpha(\beta + \gamma) = \alpha\beta + \alpha\gamma)$; we simply pick three sets, a, b, c, the latter two disjoint, and observe that $a \times (b \cup c) = (a \times b) \cup (a \times c)$.

More generally, let $(a_i)_{i \in I}$, $(b_i)_{i \in I}$ be families of sets each indexed by I. Suppose that for every $i \in I$ we have $\bar{a}_i = \bar{b}_i$. Then $\bigtimes_{i \in I} a_i$ and $\bigtimes_{i \in I} b_i$ are equipollent. This is proved by noting that for any $i \in I$ the set B_i of bijections from a_i on to b_i is non-empty; by AC'', therefore, the product of the family $(B_i)_{i \in I}$ is not empty. So there is a family $(\varphi_i)_{i \in I}$ of maps, each φ_i being a bijection from a_i on to b_i. We now define the required bijection from $\bigtimes_{i \in I} a_i$ on to $\bigtimes_{i \in I} b_i$ by associating each element $(x_i)_{i \in I}$ of the former set with $(\varphi_i(x_i))_{i \in I}$ in the latter.

If $(\alpha_i)_{i \in I}$ is a family of cardinals we call the cardinal of the set $\bigtimes_{i \in I} \alpha_i$ the *product of the family*, and write it $\prod_{i \in I} \alpha_i$. It is easily checked that this cardinal is the cardinal of the Cartesian product of any family $(a_i)_{i \in I}$ indexed by I and satisfying $\bar{a}_i = \alpha_i$ for all $i \in I$.

Similar considerations for addition lead us to the following. Let $(a_i)_{i \in I}$ and $(b_i)_{i \in I}$ be families as above which satisfy the further condition that $i \neq j \rightarrow a_i \cap a_j = b_i \cap b_j = \emptyset$. Then $\bigcup_{i \in I} a_i$ and $\bigcup_{i \in I} b_i$ are equipollent. As before, we prove this with the axiom of choice, which supplies us with a family $(\varphi_i)_{i \in I}$ of maps, each φ_i being a bijection of a_i on to b_i. Then if $x \in \bigcup_{i \in I} a_i$ it can only be in one of the a_i, say $a_{i(x)}$, and so we can couple it with $\varphi_{i(x)}(x)$ in $\bigcup_{i \in I} b_i$, thus establishing the desired correspondence between $\bigcup_{i \in I} a_i$ and $\bigcup_{i \in I} b_i$.

Given any family of cardinals $(\alpha_i)_{i \in I}$ we can construct a family of disjoint sets $(a_i)_{i \in I}$ which are equipollent to them for every $i \in I$, by writing $a_i = \alpha_i \times \{i\}$. The cardinal of $\bigcup_{i \in I} a_i$, which is quite independent of the choice of the disjoint family, is called the *sum of the family* $(\alpha_i)_{i \in I}$, and written $\sum_{i \in I} \alpha_i$.

To define exponentiation of cardinals, let $\bar{a} = \bar{\bar{a}}'$ and $\bar{b} = \bar{b}'$. Then ${}^{b}a$ (the set of maps from b into a) is equipollent with ${}^{b'}a'$. For suppose $f: a \rightarrow a'$ and $g: b \rightarrow b'$ are appropriate bijections. Then to each $\varphi \in {}^{b}a$ we can associate the map $\psi \in {}^{b'}a'$ defined by

$$\psi = f \circ \varphi \circ g^{-1}.$$

This clearly establishes a bijection between the two sets of maps.

So if α, β are cardinals, the cardinality of the set ${}^{\beta}\alpha$ is called the *βth power of* α, and written α^{β}. If a, b have cardinalities α, β respectively, it is easily checked that ${}^{b}a$ has cardinality α^{β}.

The usual laws of exponentiation for cardinals, that $\alpha^{\beta+\gamma} = \alpha^{\beta} . \alpha^{\gamma}$ and that $(\alpha^{\beta})^{\gamma} = \alpha^{\beta\gamma}$ are easily established from this definition.

FINITE ORDINALS

(This section does not use the axiom of choice.)

We recall that the next greatest ordinal after a given ordinal α is its successor $\alpha \cup \{\alpha\}$ (or $\alpha + 1$ for short; but this $+$ is different from the $+$ of cardinal addition above). If β is the successor of α we say that α is the *predecessor* of β.

An ordinal α is called *finite* if every $\beta \leqslant \alpha$ (apart from $\emptyset$) has a predecessor. The formula 'α is a finite ordinal' is easily expanded into

$On(\alpha) \wedge \forall\beta[On(\beta) \wedge \beta \subset \alpha \wedge \beta \neq \emptyset \rightarrow \exists\gamma(\beta = \gamma \cup \{\gamma\})]$. Finite ordinals are also known as *natural numbers*. It should be obvious that if α is a finite ordinal and $\beta \leqslant \alpha$, then β is finite; and that if α is finite, so is $\alpha+1$.

Mathematical Induction: Let P be a class such that $P(0)$ is true, and for every finite ordinal α, $P(\alpha) \rightarrow P(\alpha+1)$ is also true. Then $P(\alpha)$ is true for every finite ordinal α.

For if not, take the smallest finite ordinal α_0 not in P. Since 0 is in P, $\alpha_0 \neq 0$. Thus it has a predecessor β_0. As α_0 was the smallest finite ordinal not in P, we must have $P(\beta_0)$. We are given that $P(\beta_0) \rightarrow P(\beta_0+1)$. So $P(\alpha_0)$, which is impossible.

It is worth pointing out again that the definitions above of the words 'finite' and 'natural number' are not intended to provide for them their usual senses. In future, therefore, everyday uses of the words will be explicitly recorded as such.

It is easily seen that what we would intuitively say was an ordinal α with a finite number of elements is indeed finite according to the above definition. But it could happen that a finite ordinal β had, intuitively speaking, infinitely many elements; in this case since $\beta \subset \alpha$ is ruled out, β would be greater than α. In addition, the part of the universe made up of these finite ordinals (in an intuitive sense) could not then be a class; for supposing $P(x)$ to define the class, we would have $\neg P(\beta)$, and so could prove the existence of a least ordinal β_0 for which $\neg P(\beta_0)$. Then if β_0 were equal to γ_0+1, we would have $P(\gamma_0)$, showing that, intuitively speaking, γ_0 had finitely many elements; but with $\beta_0 = \gamma_0 \cup \{\gamma_0\}$ this would hardly be possible.

THEOREM: *Every finite ordinal is a cardinal.*

PROOF: This is proved by induction. 0 is obviously a cardinal. So suppose that α is a finite ordinal which is a cardinal, whilst $\alpha+1 = \alpha \cup \{\alpha\}$ is not. Then there must exist an ordinal $\gamma < \alpha+1$ and a bijection f from $\alpha+1$ on to γ. As $\alpha+1 \neq \emptyset$, γ cannot be 0; thus it is $\beta \cup \{\beta\}$ for some β, and since $\gamma < \alpha+1$, we get $\beta < \alpha$.

Were $f(\alpha)$ equal to β, $f \restriction \alpha$ would map α one-one on to β, which would contradict the fact that α is a cardinal. So $f(\alpha) = \xi \neq \beta$, and as f is bijective $f(\eta) = \beta$ for some $\eta \in \alpha+1$; $\eta \neq \alpha$, so $\eta \in \alpha$. We now define a map g which *does* map α one-one on to β, by setting $g(x)=f(x)$ for all x in α except for η; and $g(\eta)=\xi$. Since α is a cardinal, we have a contradiction. ▮

We call a cardinal *finite* if it is a finite ordinal. Let us note the following consequence of the theorem above. If α is a finite ordinal, then its successor $\alpha+1$ is indeed the cardinal sum of α and 1. Thus within the domain of finite ordinals our notation is unambiguous.

THEOREM: *If α, β are finite cardinals, so too are $\alpha+\beta$, $\alpha\beta$, α^β.*
PROOF: Each proof is done by induction on β. It is trivial that if α is finite and β is 0 then $\alpha+\beta$ is finite. Moreover, since $\alpha+(\beta+1)=(\alpha+\beta)+1$, a finite $\alpha+\beta$ means a finite $\alpha+(\beta+1)$. This proves the first part.

The proof for multiplication is similar. Since $\alpha(\beta+1)=\alpha\beta+\alpha$, the assumption that $\alpha\beta$ is finite when α, β are, together with the previous result, yields at once that $\alpha(\beta+1)$ is finite. Again, the identity $\alpha^{\beta+1}=\alpha^\beta\cdot\alpha$ supplies all that is needed to establish that α^β is finite. ▮

We are now in a position to state the one remaining axiom of Zermelo/Fraenkel set theory.

5. AXIOM OF INFINITY

This axiom says, in effect, that *there is an ordinal which is not finite.*

On the strength of the axiom we write ω (or, in some cases, N) for the first ordinal which is not finite. Thus *ω is the set of finite ordinals.* For if α is a finite ordinal we do not have $\omega\leqslant\alpha$ (which would make ω finite); so, necessarily, $\alpha<\omega$ or, equivalently, $\alpha\in\omega$. Conversely, if $\gamma\in\omega$ we have $\gamma<\omega$, and so γ is finite, by the definition of ω.

Thus the axiom of infinity can be stated in the form: *the class of finite ordinals is a set.* For the class in question is an initial segment of *On* and so, if a set, is an ordinal ω; but it is not a finite one, since no ordinal belongs to itself, and in particular $\omega\notin\omega$.

An ordinal $\alpha\neq 0$ without a predecessor is called a *limit ordinal*; it is thus a non-zero ordinal for which $\beta<\alpha\rightarrow\beta+1<\alpha$. Alternatively, a non-zero ordinal α is a limit if and only if $\alpha=\sup_{\beta<\alpha}\beta=\bigcup_{\beta<\alpha}\beta$.

For if α is not a limit it is $\gamma+1$, which means that $\sup_{\beta<\alpha}\beta=\gamma$. If it is a limit, put $\sup_{\beta<\alpha}\beta=\gamma$; since $\gamma\leqslant\alpha$ anyway, suppose $\gamma<\alpha$. By the definition of limit, then, $\gamma+1<\alpha$, and so γ is not the supposed supremum after all, since $\gamma+1 \nleq \gamma$.

So another way of formulating the axiom of infinity is: *there is a limit ordinal.*

For, given the axiom, ω could only have a predecessor if that predecessor were finite; which would make ω finite too, directly contrary to its definition. Since a limit ordinal is necessarily not finite, the converse holds too.

A fourth way of expressing the axiom of infinity is by

$$\exists x\,[0 \in x \wedge \forall y\,(y \in x \rightarrow y \cup \{y\} \in x)].$$

Given the axiom we can take ω itself for x. Conversely, suppose the above sentence holds, and choose any a satisfying it. Let α be the least ordinal not in a. It cannot be 0, and if $\alpha=\beta \cup \{\beta\}$ we get $\beta<\alpha$; thus $\beta \in \alpha$, so by the characterization of a, $\beta \cup \{\beta\} \in a$, a contradiction. Thus α is not a successor either, so it must be a limit, which is all we need to prove.

INFINITE SETS AND INFINITE CARDINALS

(Unless otherwise stated, the axiom of choice will be allowed in this section.)

A set a is called *finite* if its cardinal is finite, *infinite* otherwise; if $\bar{a} \leqslant \omega$, we say that a is *denumerable.*

Every infinite set includes an infinite denumerable subset (one, that is, equipollent with ω*).*

For if a is infinite and f a bijection of $\bar{a}$ on to a, the range of $f \restriction \omega$ is a subset of a equipollent to ω.

THEOREM: *A set a is infinite iff it is equipollent with one of its proper subsets.*
PROOF: If a is finite, and b a proper subset, take $x \in a \setminus b$. The cardinality of a is certainly not zero, but it is finite, so of the form $\alpha+1=\alpha \cup \{\alpha\}$. There is no difficulty in finding some bijection f from a on to $\alpha \cup \{\alpha\}$ which maps x to α; in this case $f \restriction b$ maps b one-one into α. Consequently $\bar{b} \leqslant \alpha$, so $\bar{b}<\bar{a}$, and b is not equipollent with a.

If a is infinite, $\bar{a} \geqslant \omega$. Writing f for some bijection from a on to $\bar{a}$, we define an injection g from a into itself by setting

$$g(x)=x, \quad \text{if} \quad f(x) \geqslant \omega;$$
$$g(x)=y, \quad \text{if} \quad f(x)<\omega \quad \text{and} \quad f(y)=f(x)+1.$$

The range of g is a proper subset of a since it does not contain $f^{-1}(0)$; thus g is a bijection from a on to a proper subset. ▌

We will write *Card'* for the class of infinite cardinals; it is a subclass of *On*, but not a set (for then *Card* would be too, since ω, the class of finite cardinals, is a set). So *Card'* is just a well-ordered class, and therefore there must be a functional relation $y = \aleph(\alpha)$ establishing an isomorphism between *On* and *Card'*. For simplicity we write the infinite cardinal $\aleph(\alpha)$ as $\aleph_\alpha$; the relation $y = \aleph_\alpha$ then abbreviates 'y is an infinite cardinal, and the set of infinite cardinals that are less than y is isomorphic, as a well-ordered set, to α'.

We have $\aleph_0 = \omega$; and, for each α, $\aleph_{\alpha+1}$ is the first cardinal $> \aleph_\alpha$.

If α is a limit ordinal, $\aleph_\alpha = \bigcup_{\beta<\alpha} \aleph_\beta$.

To show that $\bigcup_{\beta<\alpha} \aleph_\beta$ is a cardinal, suppose that $\gamma < \bigcup_{\beta<\alpha} \aleph_\beta$, so $\gamma \in \bigcup_{\beta<\alpha} \aleph_\beta$, so $\gamma \in \aleph_\beta$ for some $\beta < \alpha$. Thus $\gamma \subset \aleph_\beta$, giving

$$\bar{\bar{\gamma}} \leqslant \aleph_\beta < \aleph_{\beta+1} \subset \bigcup_{\beta<\alpha} \aleph_\beta,$$

since α is a limit ordinal.

This proves that $\overline{\overline{\bigcup_{\beta<\alpha} \aleph_\beta}} > \bar{\bar{\gamma}}$ and that there is no bijection from $\bigcup_{\beta<\alpha} \aleph_\beta$ on to any smaller ordinal γ.

Since this cardinal $\bigcup_{\beta<\alpha} \aleph_\beta \geqslant \aleph_\beta$ for all $\beta < \alpha$, it is also $\geqslant \aleph_\alpha$. But conversely $\aleph_\beta \subset \aleph_\alpha$ for all $\beta < \alpha$, so $\bigcup_{\beta<\alpha} \aleph_\beta \subset \aleph_\alpha$, and this proves the result.

By Cantor's theorem we have $2^{\aleph_\alpha} \geqslant \aleph_{\alpha+1}$, since $2^{\aleph_\alpha}$ is the cardinal of $\mathscr{P}(\aleph_\alpha)$.

The *continuum hypothesis (CH)* is the sentence $2^{\aleph_0} = \aleph_1$; alternatively, *$\mathscr{P}(\omega)$ can be so well ordered that every strict initial segment is denumerable.*

To show that this second formulation entails the first (the converse is clear), note that the ordinal of such a well-ordering could not in any case be denumerable, since, by Cantor's theorem, $\mathscr{P}(\omega)$ is not. By the specification of the well-ordering, it follows that the ordinal in question is the first non-denumerable ordinal, that is $\aleph_1$.

The *generalized continuum hypothesis (GCH)* is the sentence: $2^{\aleph_\alpha} = \aleph_{\alpha+1}$ *for every ordinal α.*

Write On^2 for the class of all ordered pairs of ordinals $\langle\alpha, \beta\rangle$. On this class we can define a well-ordering relation R as follows:

$$\begin{aligned}&\langle\alpha, \beta\rangle \leqslant \langle\gamma, \delta\rangle \text{ (mod. } R) \text{ if and only if}\\ &\sup(\alpha, \beta) < \sup(\gamma, \delta)\\ &\vee \sup(\alpha, \beta) = \sup(\gamma, \delta) \wedge \alpha < \gamma\\ &\vee \sup(\alpha, \beta) = \sup(\gamma, \delta) \wedge \alpha = \gamma \wedge \beta \leqslant \delta .\end{aligned}$$

Every segment is seen to be a set. For if $S_\xi(R)$ is an initial segment, and $\xi = \langle\alpha_0, \beta_0\rangle$, and $\langle\alpha, \beta\rangle < \langle\alpha_0, \beta_0\rangle$ (mod. R), then

$$\sup(\alpha, \beta) \leqslant \sup(\alpha_0, \beta_0).$$

Putting $\gamma_0 = \sup(\alpha_0, \beta_0)$, we note that every pair $\langle\alpha, \beta\rangle$ in $S_\xi(R)$ is a member of the set $(\gamma_0 + 1)^2$; so $S_\xi(R)$ is a subclass of that set, and thus a set itself.

On the other hand, take a non-empty set X all of whose elements are in On^2; the set of all $\sup(\alpha, \beta)$ for $\langle\alpha, \beta\rangle \in X$ has a smallest element γ_0; the set of all α such that for some β we have both $\langle\alpha, \beta\rangle \in X$ and $\sup(\alpha, \beta) = \gamma_0$ has a smallest element α_0; and the set of β for which $\sup(\alpha_0, \beta) = \gamma_0$ and $\langle\alpha_0, \beta\rangle \in X$ has a least element β_0.

From all this it drops out that $\langle\alpha_0, \beta_0\rangle$ is the least element of X (mod. R); and so R is a well-ordering.

Since On^2 is therefore a well-ordered class, but not a set, a relation $\gamma = J(\alpha, \beta)$ can be found which establishes an isomorphism between On^2 and On. In the definition of J, be it noted, there is no need for the axiom of choice.

THEOREM: *For each ordinal* α, $\aleph_\alpha^2 = \aleph_\alpha$.

PROOF: If the theorem is false, it fails first at some ordinal ρ. To obtain a contradiction we need only show that $\aleph_\rho \cdot \aleph_\rho \leqslant \aleph_\rho$, since there is no trouble with $\aleph_\rho^2 \geqslant \aleph_\rho$. Restricting J to $\aleph_\rho \times \aleph_\rho$ we get a one-one map, and so the proof reduces to showing that for $\alpha, \beta \in \aleph_\rho$, $J(\alpha, \beta) \in \aleph_\rho$.

Now $J(\alpha, \beta) \in \aleph_\rho \leftrightarrow J(\alpha, \beta) < \aleph_\rho$, so, $\aleph_\rho$ being a cardinal,

$$J(\alpha, \beta) \in \aleph_\rho \leftrightarrow \overline{\overline{J(\alpha, \beta)}} < \aleph_\rho .$$

So if we prove that $\overline{\overline{J(\alpha, \beta)}} < \aleph_\rho$ for $\alpha, \beta \in \aleph_\rho$, we are finished.

Since $J(\alpha, \beta)$ is the set of ordinals less than $J(\alpha, \beta)$, it is the range of J restricted to the set of all $\langle\alpha', \beta'\rangle$ less than $\langle\alpha, \beta\rangle$ (mod. R). Writing γ

for $\sup(\alpha, \beta)$, we easily see that this latter set is included in $(\gamma+1)^2$ (since $\langle\alpha', \beta'\rangle < \langle\alpha, \beta\rangle \to \alpha' < \gamma+1 \wedge \beta' < \gamma+1$). Thus its cardinal cannot exceed the cardinal of $(\gamma+1)^2$; but as $\gamma < \aleph_\rho$, so is $\gamma+1$, and therefore by the definition of ρ there are just two cases: $\gamma+1$ is finite, or $\overline{\overline{(\gamma+1)^2}} = \overline{\overline{\gamma+1}} < \aleph_\rho$.

Each case separately yields $\overline{\overline{J(\alpha, \beta)}} < \aleph_\rho$, and so the theorem is proved. ∎

COROLLARY: *$\aleph_\alpha^n = \aleph_\alpha$ for every natural number $n \geqslant 1$, and every ordinal α.*
PROOF: Immediate, by induction on n. ∎

COROLLARY: $\aleph_\alpha \cdot \aleph_\beta = \sup(\aleph_\alpha, \aleph_\beta)$.
PROOF: If $\aleph_\alpha < \aleph_\beta$, then clearly

$$\aleph_\beta = 1 \cdot \aleph_\beta \leqslant \aleph_\alpha \cdot \aleph_\beta \leqslant \aleph_\beta \cdot \aleph_\beta = \aleph_\beta,$$

and so $\aleph_\alpha \cdot \aleph_\beta = \aleph_\beta$, as required. ∎

COROLLARY: *Let $(\kappa_i)_{i \in I}$ be a family of cardinals, all non-zero. If $\overline{\overline{I}}$, or any κ_i, is infinite, then $\sum_{i \in I} \kappa_i = \sup(\overline{\overline{I}}, \sup_{i \in I} \kappa_i)$.*
PROOF: Recall that $\sum_{i \in I} \kappa_i$ is defined as the cardinal of the set $X = \bigcup_{i \in I} \kappa_i \times \{i\}$. Since each $\kappa_i \neq 0$, we have that $X \supset \bigcup_{i \in I} \{0\} \times \{i\}$, that is that X includes $\{0\} \times I$, which is equipollent with I. Thus $\overline{\overline{X}} \geqslant \overline{\overline{I}}$. It is clear that $\overline{\overline{X}} \geqslant \kappa_i$ for every $i \in I$, and so $\overline{\overline{X}} \geqslant \sup(\overline{\overline{I}}, \sup_{i \in I} \kappa_i)$.

For the reverse inequality, put $\kappa = \sup_{i \in I} \kappa_i$. Then $X \subset \bigcup_{i \in I} \kappa \times \{i\} = \kappa \times I$. Thus $\overline{\overline{X}} \leqslant \kappa \cdot \overline{\overline{I}}$ which, since either κ or $\overline{\overline{I}}$ is infinite, is equal to $\sup(\overline{\overline{I}}, \kappa)$. ∎

An isomorphism between the class of finite sequences of ordinals and On: (To establish this we do not need the axiom of choice.)

For a class C, a *finite sequence* s of objects of C is, by definition, a map whose domain is a natural number, and whose values are in C. This natural number, the domain, is otherwise called the *length* of the sequence, and is written $l(s)$. We write $\sigma(C)$ for the class of all finite sequences of objects of C.

An ordering R can be defined on $\sigma(On)$, the class of finite sequences of ordinals, by setting

$$\begin{aligned} & s < t(\text{mod. } R) \quad \text{if and only if} \\ & \sup(s) < \sup(t) \vee \\ & \sup(s) = \sup(t) \wedge\ l(s) < l(t) \vee \\ & \sup(s) = \sup(t) \wedge\ l(s) = l(t) \wedge s \neq t \wedge \\ & \qquad s(n) < t(n) \text{ for the first } n \text{ where} \\ & \qquad s(n) \neq t(n). \end{aligned}$$

(Here $\sup(s)$ stands for the greatest value taken by the function s.)

We show that R, defined in this way, is a well-ordering. This involves, in the first place, that the class $S_t(R)$ of sequences $<t$ (mod. R) be a set; but since everything in this class is also in the set $\sigma(\gamma+1)$ for $\gamma=\sup(t)$, this is easily proved.

So let X be a non-empty set of finite sequences of ordinals. As s ranges over X, $\sup(s)$ takes on various ordinals as values, and of these there must be a least, γ say. Amongst the sequences s in X for which $\sup(s)=\gamma$, there will be at least one of minimal length, of length n say. Now define recursively a map f with domain n as follows: if $i \in n$, $f(i)=$ the smallest ordinal of the form $s(i)$ for $s \in X$ satisfying $\sup(s)=\gamma$, $l(s)=n$, and $s(j)=f(j)$ for every $j<i$.

Now f, as thus defined, is a member of X; and, further, $\sup(f)=\gamma$ and $l(f)=n$. For it is immediate, by induction on i, that for every $i \leqslant n$ there is an $s \in X$ for which $\sup(s)=\gamma$, $l(s)=n$, and $s \restriction i=f \restriction i$; whence, putting $i=n$, the result. Moreover, f is the smallest element of X. For take some s in X, distinct from f. If $\sup(s) \neq \gamma$ we must have $\sup(s)>\gamma=\sup(f)$, by the definition of γ. Thus $s>f$. If $\sup(s)=\gamma$, however, and $l(s) \neq n$, we have in the same way $l(s)>n=l(f)$; and so $s>f$. Finally, if $\sup(s)=\gamma$ and $l(s)=n$, let $i<n$ be the first integer where $s(i) \neq f(i)$. By the definition of f, $s(i)>f(i)$, so once more $s>f$.

But a well-ordered class that is not a set is isomorphic to On; and there is consequently a functional relation $s=J(\alpha)$ which establishes an isomorphism between On and $\sigma(On)$.

CHAPTER III

THE AXIOM OF FOUNDATION

The axiom of foundation (*AF* for short) is the sentence

$$\forall x[x \neq \emptyset \rightarrow \exists y(y \in x \wedge y \cap x = \emptyset)];$$

in other words, the axiom states that every non-empty set has an element which is disjoint from it.

From the axiom it follows that there is no infinite sequence (that is, no map with domain ω) $(u_n)_{n\in\omega}$ for which $u_{n+1} \in u_n$ for all $n \in \omega$; for if x were the set of terms of such an infinite descending sequence (its range, in other words), it would fail to contain an element disjoint from itself.

In particular, therefore, the axiom implies that for every x, $x \notin x$; and, more generally, that there are no $\in$-cycles – finite sequences $x_0, \ldots, x_{n-1}$ that is, satisfying $x_0 \in x_1 \in \ldots \in x_{n-1} \in x_0$. For, again, the set $\{x_0, \ldots, x_{n-1}\}$ would contradict *AF*.

We can take advantage of the axiom to provide a more simple definition of ordinals than the one given earlier. This is established in the following theorem.

THEOREM: *For X to be an ordinal it is necessary and sufficient that the two conditions*

$$\forall u \forall v[u \in X \wedge v \in X \rightarrow (u \in v \vee u = v \vee v \in u)]$$

and

$$\forall u[u \in X \rightarrow u \subset X]$$

be satisfied.

PROOF: The necessity of both conditions is immediate. Conversely, we note first that $\in$ imposes a strict linear order on any X satisfying them. For by *AF* we cannot have $u \in v$ and $v \in u$ together; moreover, when u v, $w \in X$, and $u \in v$ and $v \in w$, the possibility $u=w$ is ruled out (for then $u \in v$ and $v \in u$), and the possibility $w \in u$ is ruled out (for then we get the cycle $u \in v \in w \in u$); and so by the first condition all that is left is $u \in w$.

To show that this ordering is a well-ordering, we take any non-empty $Y \subset X$; by *AF* there is some $u \in Y$ satisfying $u \cap Y = \emptyset$. But this means that if $v \in Y$ then $v \notin u$, and therefore that u is the desired least element (mod. $\in$) of Y. ∎

Without necessarily supposing the axiom of foundation to hold in the universe $\mathscr{U}$, we can define by recursion a functional relation $y = V_\alpha$ by setting

$$V_0 = \emptyset;$$
$$V_\alpha = \bigcup_{\beta<\alpha} \mathscr{P}(V_\beta).$$

It is clear enough from this that $\alpha \leqslant \alpha' \to V_\alpha \subset V_{\alpha'}$. Consequently, since $V_{\alpha+1} = \bigcup_{\beta\leqslant\alpha} \mathscr{P}(V_\beta)$, we have $V_{\alpha+1} = \mathscr{P}(V_\alpha)$ *for every ordinal* α.

On the other hand, for limit α we have $\beta<\alpha \to \beta+1<\alpha$, so $V_\alpha = \bigcup_{\beta<\alpha} \mathscr{P}(V_\beta) = \bigcup_{\beta<\alpha} V_{\beta+1} = \bigcup_{\beta<\alpha} V_\beta$ *for every limit ordinal* α.

The class of all elements of some V_α or another is called V; it is defined by the formula $V(x)\colon \exists\alpha(On(\alpha) \wedge x \in V_\alpha)$.

For any x in V we define the *rank* of x, $\mathrm{rn}(x)$, as the least ordinal α for which $x \in V_\alpha$. Since for limit α we have $x \in V_\alpha \to x \in \bigcup_{\beta<\alpha} V_\beta$, $\mathrm{rn}(x)$ is always a successor, of the form $\beta+1$.

LEMMA: *A set a is in V iff all its elements are; if $\mathrm{rn}(a) = \alpha$, then every element of a has rank $<\alpha$.*

PROOF: Suppose a is in V, and its rank is $\alpha = \beta+1$. Then $a \in V_{\beta+1}$, so $a \subset V_\beta$. Thus all a's elements are in V; moreover, their ranks are all $\leqslant\beta$, and so $<\alpha$. Conversely, suppose that every element of x is in V. The ranks of these elements form a set, and so must be bounded above by some ordinal α. It follows that $a \subset V_\alpha$, whence $a \in V_{\alpha+1}$; thus a is in V. ∎

THEOREM: *Every ordinal α is in V, and the rank of α is $\alpha+1$.*

PROOF: If there is one, let α be the first ordinal for which $\alpha \notin V_{\alpha+1}$; then $\beta \in \alpha \to \beta \in V_{\beta+1}$, and so

$$\alpha \subset \bigcup_{\beta<\alpha} V_{\beta+1} = \bigcup_{\beta<\alpha} \mathscr{P}(V_\beta) = V_\alpha,$$

from which $\alpha \in V_{\alpha+1}$, contrary to hypothesis.

Likewise, let α be the first ordinal, if there is one, such that $\alpha \in V_\alpha$. Then $\alpha \in \bigcup_{\beta<\alpha} \mathscr{P}(V_\beta)$, so α is in $\mathscr{P}(V_\beta)$ for some $\beta<\alpha$. But then $\alpha \subset V_\beta$, and since $\beta \in \alpha$ we obtain $\beta \in V_\beta$, in contradiction to the definition of α. ∎

THEOREM: *The axiom of foundation holds if and only if V is the whole universe; or, briefly,* $AF \leftrightarrow \forall x V(x)$.

PROOF: Supposing $\forall x V(x)$ to hold, we pick a non-empty set a, and from its elements pick some set b of minimum rank α. Then b's elements all have rank $<\alpha$, and so cannot themselves be in a. Thus $b \cap a = \emptyset$, as required for AF.

The converse is most easily proved by defining the transitive closure of a set as follows. Recall that a set X is transitive if $\forall x \forall y [x \in X \wedge y \in x \rightarrow \rightarrow y \in X]$. The *transitive closure* of X is defined as the most restricted (that is, smallest under the order $\subset$) set $\mathscr{C}(X)$ which is transitive and includes X. That this definition is a correct one is established by the theorem which we interpose here.

THEOREM: *For any set X there exists a unique transitive set Y which includes X and is itself included in every transitive set including X.*

PROOF: We define a map f on ω by recursion: $f(0) = X$; $f(n+1) = \bigcup_{x \in f(n)} x$. Let $Y = \bigcup_{n \in \omega} f(n)$. Then it is obvious that $X \subset Y$; and further, Y is transitive. For if $x \in Y$, then $x \in f(n)$ for some $n \in \omega$, so $x \subset f(n+1)$, so $x \subset Y$. Now if Z is a transitive superset of X, we prove by induction that every $f(n) \subset Z$; $n=0$ is immediate; if $x \in f(n) \subset Z$, $x \in Z$, so $x \subset Z$, because Z is transitive; therefore $f(n+1) = \bigcup_{x \in f(n)} x \subset Z$ as required. This completes the inductive step, and we can conclude that

$$Y = \bigcup_{n \in \omega} f(n) \subset Z. \blacksquare$$

Returning to the stranded theorem, suppose that AF holds, but that some object a is not in V. Let b be a transitive set including a, and b' the set of elements of b that are not in V.

Then b' is non-empty; for a is not in V, so has an element x_0 not in V; and since $a \subset b$, x_0 is also in b'.

But for every $x \in b'$ we also have $b' \cap x$ non-empty. For if $x \in b'$, it is not in V, so has an element y not in V. Since $y \in x \in b$, and b is transitive, $y \in b$; and thus $y \in b'$, by the definition of b'. But then $y \in b' \cap x$, which therefore cannot be empty.

Thus b' contradicts AF, and the theorem is established. $\blacksquare$

Let us note one property of transitive closures.

THEOREM: $\mathscr{C}(X)=X \cup \bigcup_{y \in X} \mathscr{C}(y)$.

PROOF: $\mathscr{C}(X)$ certainly includes X. If $y \in X$ then $y \subset \mathscr{C}(X)$, and as $\mathscr{C}(y)$ is included in every transitive set including y, $\mathscr{C}(y) \subset \mathscr{C}(X)$. Thus $X \cup \bigcup_{y \in X} \mathscr{C}(y) \subset \mathscr{C}(X)$. Conversely, if $Z=X \cup \bigcup_{y \in X} \mathscr{C}(y)$, then $X \subset Z$, and, moreover, Z is transitive. For if $v \in u \in Z$, then either u is in some $\mathscr{C}(y)$ with $y \in X$, in which case $v \in \mathscr{C}(y)$; or $u \in X$, in which case $v \in u \subset \mathscr{C}(u)$. Either way, $v \in \bigcup_{y \in X} \mathscr{C}(y)$, so $v \in Z$, as desired. Thus $X \subset Z$, and Z is transitive; so $\mathscr{C}(X) \subset Z$, and the proof is complete. ∎

Well-founded Relations and Extensional Relations: Let r be a binary relation on a set a (so $r \subset a^2$); r is said to be *well founded on a* if for every non-empty $X \subset a$ there exists an $x \in X$ such that $\langle y, x\rangle \in r$ holds for no $y \in X$. It should be clear that if $b \subset a$, the restriction $r \cap b^2$ of r to b is well founded on b.

A map φ with domain a is called *collapsing for r* if for all $x \in a$ we have

$$\varphi(x) = \{\varphi(y) \mid y \in a \wedge \langle y, x\rangle \in r\}.$$

Evidently, if φ is collapsing the range of φ is a transitive set, since each element of $\varphi(x)$, for $x \in a$, is of the form $\varphi(y)$.

THEOREM: *If r is a well-founded relation on a there exists one and only one map with domain a which collapses for r. Moreover, the range of this map is in the class V.*

PROOF: (*Uniqueness*) Let φ, ψ be two such collapsing maps, and write X for $\{x \in a \mid \varphi(x) \neq \psi(x)\}$. If $\varphi \neq \psi$, X must be non-empty, so there will be an $x_0 \in X$ for which $\forall y \in X(\langle y, x_0\rangle \notin r)$. Since $x_0 \in X$, we have $\varphi(x_0) \neq \psi(x_0)$, and so without loss of generality we can assume a u that is in $\varphi(x_0)$ but not in $\psi(x_0)$. Now φ is collapsing, so there is a $y \in a$ for which $\varphi(y)=u$ and $\langle y, x_0\rangle \in r$. By the specification of x_0, y cannot be in X; so $\varphi(y)=\psi(y)=u$. But ψ collapses too, so from $\psi(y)=u$ and $\langle y, x_0\rangle \in r$ we get $u \in \psi(x_0)$, which is a contradiction.

(*Existence*) A subset Z of a is called *r-transitive* if $x \in Z \wedge \langle y, x\rangle \in r \to \to y \in Z$. Clearly if φ has domain a and is collapsing for r then $\varphi \restriction Z$ is collapsing for $r \cap Z^2$ when Z is an r-transitive subset of a. If Z, Z' are both r-transitive subsets of a, so too is $Z \cap Z'$; and so if Φ, Φ' are collapsing maps with domains Z, Z' respectively, $\Phi \upharpoonright Z \cap Z'$ and $\Phi' \upharpoonright Z \cap Z'$ are also collapsing maps, both with domain $Z \cap Z'$; and so, by the uniqueness part of the theorem, they are identical.

So let φ be the union of all the collapsing maps Φ that are defined on any r-transitive subset of a. From our discussion above it follows that φ is also a map, and that its domain Y is also an r-transitive subset of a (since it is the union of a set of r-transitive subsets).

Now φ is obviously collapsing; and so Y must be the largest r-transitive subset of a which is the domain of a collapsing map.

Is Y equal to a itself? If not, take some $x_0 \in a \setminus Y$ such that $\forall y \in a \setminus Y(\langle y, x_0\rangle \notin r)$; that such an x_0 would exist under these conditions follows at once from the well-foundedness of r on a. We now define a map ψ on $Y \cup \{x_0\}$ such that ψ and φ are identical on Y and $\psi(x_0) = \{\varphi(y) \mid y \in a \wedge \langle y, x_0\rangle \in r\}$ (this definition is a correct one since $\langle y, x_0\rangle \in r \rightarrow y \in Y$).

But now it falls out at once that $Y \cup \{x_0\}$ is an r-transitive subset of a and that ψ is collapsing. And this is impossible, since the domain of ψ is $Y \cup \{x_0\}$, which properly includes Y.

Consequently $Y = a$; and φ is a collapsing map with domain a.

It remains to show that $rng(\varphi)$ is in V, or, what is the same thing, that all its elements y are. Suppose otherwise, and put $X = \{x \in a \mid \neg V(\varphi(x))\}$; X is then non-empty. As r is well founded there must be an $x_0 \in X$ such that $\forall y \in X(\langle y, x_0\rangle \notin r)$. Thus $\neg V(\varphi(x_0))$, which means that there is a $u \in \varphi(x_0)$ such that $\neg V(u)$.

On the other hand, as φ is collapsing, there exists a $y \in a$ such that $\varphi(y) = u$ and $\langle y, x_0\rangle \in r$. Thus $y \notin X$, in view of the choice of x_0; so $V(\varphi(y))$, so $V(u)$; and this contradicts the conclusion of the previous paragraph. ▌

A binary relation r on a set a is called *extensional* if, whatever elements x, y may be of a, if $\forall z \in a[\langle z, x\rangle \in r \leftrightarrow \langle z, y\rangle \in r]$ then $x = y$; or, more succinctly, if the set a, together with the relation r, satisfies the axiom of extensionality.

A set a is called *extensional* if the membership relation is extensional on a; that is, if $x \cap a = y \cap a \rightarrow x = y$ for any x, y in a.

Observe that every transitive set is extensional; for if a is a transitive set and $x \in a$ then $x \cap a = x$.

THEOREM: *Let r be an extensional relation which is well founded on a. Then there exists a unique isomorphism from a, together with the binary relation r, on to a transitive set (together with the membership relation); and this transitive set is in V.*

PROOF: Any isomorphism such as that required by the theorem is evidently a collapsing map; and so by the previous theorem we have uniqueness. To prove that there is any such isomorphism at all, we consider that one map φ defined on a which collapses for r, and show that it fills the bill. Let b be the range of φ; we have already seen that b is transitive and in V.

Firstly, φ is injective. For otherwise there would be an element d of b of minimal rank for which there were two inverse images $x, y \in a$; that is to say, $x, y \in a$, $x \neq y$, and $\varphi(x) = \varphi(y) = d$. As x, y are distinct and r is extensional, we can assume, without loss of generality, that we have some $u \in a$ for which $\langle u, x \rangle \in r$ and $\langle u, y \rangle \notin r$. The former yields $\varphi(u) \in \varphi(x)$ – since φ is collapsing; so $\varphi(u) \in \varphi(y)$ too. Thus for some $v \in a$ we have $\langle v, y \rangle \in r$ and $\varphi(v) = \varphi(u)$; and as $\langle u, y \rangle \notin r$, we also have $v \neq u$. But $\varphi(u) = \varphi(v) \in d$, whence $\varphi(u)$, $\varphi(v)$ are of rank less than d, in contradiction to the definition of d.

Secondly, $\langle y, x \rangle \in r \rightarrow \varphi(y) \in \varphi(x)$, since φ is collapsing. And lastly, the converse of this is proved as follows. If $\varphi(y) \in \varphi(x)$ then there is a $z \in a$ such that $\langle z, x \rangle \in r$ and $\varphi(z) = \varphi(y)$ (since φ is a collapsing map). But φ is also injective, so $z = y$. Thus $\langle y, x \rangle \in r$.

This concludes the proof that φ is indeed an isomorphism. ▮

Note that this theorem is a generalization of the theorem on pp. 16f. For if r is a strict well-ordering on a it is easily seen to be both extensional and well founded. What is more, the range set of the collapsing isomorphism is transitive and well ordered by $\in$; an ordinal, no less.

The following corollary is proved with the help of the axiom of foundation. It will be used in Chapter VIII.

COROLLARY: *Every extensional set is isomorphic, under a unique isomorphism, to a transitive set.*

PROOF: Let a be extensional, and r the restriction of the membership relation to a; thus $r = \{\langle x, y \rangle \in a^2 \mid x \in y\}$. It follows at once from AF that r is well founded; it is extensional, by hypothesis; and so the theorem above gives the result. ▮

THE HEREDITARILY FINITE SETS

A set a is called *hereditarily finite* if it is in V_ω, or, in other words, if it is in V_n for some natural number n. Without having to use the axiom of choice we are able to define a functional relation without parameters that maps V_ω one-one on to ω.

LEMMA: *If n is a natural number, ${}^n\{0, 1\}$ is equipollent with a unique natural number.*

PROOF: The uniqueness is trivial since no natural number is equipollent with any ordinal apart from itself. The rest is a simple induction on n. Given that ${}^n\{0, 1\}$ is equipollent with k, say, ${}^{n+1}\{0, 1\}$ is equipollent with $k \times \{0, 1\}$, and thus with $2k$. And that is all. ▌

We now define recursively on the natural numbers a sequence of bijections φ_n from V_n on to a natural number v_n. For $n=0$, $\varphi_n=\emptyset$. So suppose that we are presented with φ_n: $V_n \to v_n$; our object is to define φ_{n+1}: $V_{n+1} \to v_{n+1}$.

In terms of φ_n we can define a bijection ψ_n: $\mathscr{P}(V_n) \to {}^{v_n}\{0, 1\}$ (that is, from V_{n+1} on to ${}^{v_n}\{0, 1\}$) by

$$\psi_n(X)(i) = 1 \quad \text{if} \quad \varphi_n^{-1}(i) \in X,$$
$$\psi_n(X)(i) = 0 \quad \text{if} \quad \varphi_n^{-1}(i) \notin X,$$

where $X \subset V_n$ and $0 \leqslant i < v_n$.

But ${}^{v_n}\{0, 1\}$ is a set of finite sequences of ordinals, binary sequences in fact (that is, the only ordinals appearing are 0, 1), of length v_n. It is thus well ordered by the relation $R(x, y)$ defined on $\sigma(On)$ on pp. 33f. By the lemma, the ordinal isomorphic to this well-ordering is some natural number v_{n+1}. Thus we get our bijection φ_{n+1} from V_{n+1} on to v_{n+1} by writing $\varphi_{n+1}=H(\varphi_n)$, where $\varphi'=H(\varphi)$ is the relation 'Where u is the domain of φ and v the range, φ' is that map from $\mathscr{P}(u)$ which is composed of the following two maps: the map ψ: $\mathscr{P}(u) \to {}^v\{0, 1\}$ developed from φ, and the unique isomorphism from ${}^v\{0, 1\}$ (as well ordered by R) on to its ordinal'. This relation is determined by a parameter-free formula, and so the functional relation $n \to \varphi_n$ defined on ω is also void of parameters.

So we have a parameter-free functional relation $n \to r_n$ which associates with each natural number n a well-ordering r_n on V_n (r_n is the inverse image under φ_n of the well-ordering of v_n). We can now define a well-ordering r

on V_ω by setting $x<y$ (mod. r) iff

$$\mathrm{rn}(x) < \mathrm{rn}(y) \vee [\mathrm{rn}(x) = \mathrm{rn}(y) = n \wedge x < y(\text{mod. } r_n)].$$

Every initial segment under r is isomorphic to a natural number; for the set of r-predecessors of x_0 is a subset of V_n where n is $\mathrm{rn}(x_0)$; and so its ordinal is finite.

This shows that the well-ordering r on V_ω is isomorphic to the well-ordering of ω; and the isomorphism K between ω (ordered by membership) and the well-ordered system $\langle V_\omega, r\rangle$ is the functional relation sought, since it has been defined without parameters.

RELATIVIZED FORMULAS

Let X be some class, and $E(x_0,\ldots, x_{n-1})$ some formula in (at most) the variables $x_0,\ldots, x_{n-1}$, whose parameters are all objects of X. We write $E^X(x_0,\ldots, x_{n-1})$ for the *relativization of the formula E to X*, namely that formula defined in the following way by (informal) recursion on the length of E. (Since the formulas are not objects of the universe, there is no question here of a recursion on the integers of the universe, but one on intuitive objects.)

(1) E is of the form $x\in y$, $x=y$, $x\in a$, $x=a$, $a\in x$, $a=x$, $a\in b$, $a=b$ (a, b, objects of X). E is in this case called *atomic*, and is its own relativization.

(2) E is $\neg F$; then E^X is $\neg F^X$.

(3) E is $F \vee G$; then E^X is $F^X \vee G^X$.

(4) E is $\exists x F(x, x_0,\ldots, x_{n-1})$;

then E^X is $\exists x[X(x) \wedge F^X(x, x_0,\ldots, x_{n-1})]$.

It follows at once from (2) and (4) that if E is $\forall x F(x, x_0,\ldots, x_{n-1})$ – that is, at length, $\neg\exists x \neg F(x, x_0,\ldots, x_{n-1})$ – then E^X is $\forall x[X(x) \rightarrow F^X(x, x_0, \ldots, x_{n-1})]$.

Note that if the class X qualifies as a set, then the relation $X(x)$ takes the form $x\in X$.

Consider any n objects $a_0,\ldots, a_{n-1}$ of the class X. Then it is apparent that the formula $E^X(a_0,\ldots, a_{n-1})$ is true in the universe iff the formula $E(a_0,\ldots, a_{n-1})$ is true in the class X (augmented by the binary relation $\in\restriction X$). As an example, a set X is extensional if and only if the formula E^X is true (in the universe), where E itself is one form of the axiom of extensionality,

$$\forall x \forall y [x = y \leftrightarrow \forall z (z \in x \leftrightarrow z \in y)].$$

RELATIVE CONSISTENCY OF THE AXIOM OF FOUNDATION

In this section we shall show that the deployment of the axiom of foundation cannot of itself lead to contradiction within ZF, by assuming a model for ZF – that is, a universe $\mathscr{U}$ – and from it constructing a model for $ZF+AF$; this latter model will be a set (in the intuitive sense), augmented by a binary relation, for which all the ZF axioms hold, and the axiom of foundation too. More precisely, we show that if ZF has a model, so also has $ZF+AF$; this is what is meant by *relative consistency*. In fact, *if $\mathscr{U}$ is a universe, then the ZF axioms and the axiom of foundation all hold in the class V constructed within $\mathscr{U}$.*

This amounts to proving that for each axiom A of $ZF+AF$, A^V holds in $\mathscr{U}$.

Axiom of Extensionality: If x, y are in V, so too are all their members; so if they have the same members in V, they have the same members all told. Thus $x=y$, since the axiom of extensionality holds in $\mathscr{U}$.

Union Axiom: If x is in V, anything in $\bigcup x$ is obviously also in V, so by the lemma on p. 36, $\bigcup x$ itself is in V.

Power-set Axiom: If x is in V and $y \in \mathscr{P}(x)$, then $y \subset x$, and so all y's elements are in V; thus y is too, and by the same lemma, $\mathscr{P}(x)$ belongs to V.

Scheme of Replacement: Let a be a set in V, and $R(x, y)$ some formula (whose parameters are in V) which, within the class V, defines a functional relation. Clearly this functional relation is defined in $\mathscr{U}$ by the formula $V(x) \wedge V(y) \wedge R^V(x, y)$; since replacement holds in $\mathscr{U}$, we can apply it to this functional relation, and obtain in $\mathscr{U}$ a set b made up of the images of a under it. Since everything in b is in V, b itself must be, and the proof is complete.

Axiom of Infinity: Every ordinal is in V, ω in particular. Since $\emptyset \in \omega \wedge$ $\wedge \forall x[x \in \omega \rightarrow x \cup \{x\} \in \omega]$, the axiom of infinity holds in V.

Axiom of Foundation: Take a non-empty set a in V. Let b be some element of a of minimal rank. But every element of b is of lower rank than b, and therefore cannot appear in a. Thus $b \cap a = \emptyset$.

This proof of relative consistency – like all others – can be looked at from a different angle. Without reference to models, we can treat the theory ZF as a *formal system*, in the sense that its axioms and, more generally, its parameter-free formulas are regarded as mere finite sequences of symbols. The rules of inference (which need not be spelt out here, but can be found in [4] for example) are then formal rules allowing the derivation of certain finite sequences from certain others. A *proof* is just a finite sequence of formulas $A_0, \ldots, A_{n-1}$, each of which, if not an axiom, is derivable from earlier formulas in the sequence by one of the rules of inference. A *theorem* of the theory is any sentence which is the last formula of some proof, and the theory is called *consistent* (or non-contradictory) if and only if $0 \neq 0$ is not a theorem.

The relative consistency of the axiom of foundation can then be expressed as: if $0 \neq 0$ is not a theorem of ZF, then it is not a theorem of $ZF + AF$. We can prove this straightforwardly by supposing a proof of $0 \neq 0$ in $ZF + AF$, say the sequence $A_0, \ldots, A_{n-2}, 0 \neq 0$, and constructing from it a proof of $0 \neq 0$ in ZF alone.

To do this we note that every parameter-free sentence E which we have shown to hold in any universe $\mathscr{U}$ has in effect been proved from the axioms of ZF by means of the formal rules of inference. We can thus consider the sequence of relativized formulas $A_0^V, \ldots, A_{n-2}^V, (0 \neq 0)^V$. If A_i is an axiom of ZF, we have just shown that A_i^V is a theorem of ZF. Thus this sequence furnishes a proof in ZF of $(0 \neq 0)^V$, and therefore one of $(0 \neq 0)$ itself.

We give two more simple examples of relative consistency proofs.

CONSISTENCY OF THE NEGATION OF THE AXIOM OF INFINITY

We shall show here that if ZF is consistent then so also is the theory T obtained by striking out the axiom of infinity from the list of axioms, and admitting its negation in its place. In fact, *if $\mathscr{U}$ is a universe, the set V_ω (together with the membership relation) satisfies every axiom of T.*

Since V_ω is a transitive set, the extensionality axiom automatically holds. To show the union axiom holds, take $a \in V_\omega$; for some $n \in \omega$, then, $a \in V_n$, in which case $\bigcup a \subset V_n$ for sure, so $\bigcup a \in V_{n+1}$. Thus $\bigcup a \in V_\omega$. Likewise, every subset of a is in V_{n+1}, so that $\mathscr{P}(a) \subset V_{n+1}$, which gives $\mathscr{P}(a) \in V_{n+2}$; and this shows that the power-set axiom holds.

Let a be in V_ω, and take an arbitrary formula $R(x, y)$ with parameters in V_ω, which, interpreted in V_ω, defines a functional relation. In $\mathscr{U}$, then, the functional relation can be defined as $x \in V_\omega \wedge y \in V_\omega \wedge R^{V_\omega}(x, y)$ – indeed, since its domain is a set, this functional relation is a map f with values in V_ω. Let $\varphi: a \to \omega$ assign to each element x of a the rank of $f(x)$; then since $a \subset V_n$ for some $n \in \omega$, a is equipollent with a finite ordinal, and so the values of φ are all less than some natural number k. Thus b, the set of images of members of a under f, is a subset of V_k, and so $b \in V_{k+1}$ as desired.

But the axiom of infinity fails in V_ω. For suppose that a is a set, that $0 \in a$, and that if $x \in a$ so is $x \cup \{x\}$. Then every $n \in \omega$ is in a, so that $\omega \subset a$. But an element of V_ω cannot have ω as a subset since it is bound to be equipollent to some finite ordinal. Thus no such a is in V_ω.

CONSISTENCY OF THE AXIOM OF ACCESSIBILITY

In this section we suppose that the axiom of choice holds in the universe $\mathscr{U}$.

A cardinal λ is called (strongly) *inaccessible* if it has the following three properties:

(1) $\lambda > \omega$;

(2) If β is a cardinal $< \lambda$, then $2^\beta < \lambda$;

(3) If $(\beta_i)_{i \in I}$ is a family of cardinals less than λ, indexed by a cardinal $I < \lambda$, then $\sup_{i \in I} \beta_i < \lambda$.

A cardinal that is not inaccessible is called *accessible*. Note that nothing prevents a cardinal less than an accessible cardinal from being inaccessible; for example, if λ is inaccessible, 2^λ, though strictly greater than λ, is accessible.

LEMMA: If λ *is an inaccessible cardinal then* $\overline{\overline{V_\lambda}} = \lambda$*; a set a is in* V_λ *iff* $a \subset V_\lambda$ *and* $\bar{\bar{a}} < \lambda$.

PROOF: Since $\lambda \subset V_\lambda$ we certainly have $\overline{\overline{V_\lambda}} \geqslant \lambda$. To obtain the reverse inequality we first show that if $\alpha < \lambda$ then $\overline{\overline{V_\alpha}} < \lambda$. So let α be the first ordinal $< \lambda$, if there is one, for which $\overline{\overline{V_\alpha}} \geqslant \lambda$. We have $\overline{\overline{V_\alpha}} = \overline{\overline{\bigcup_{\beta<\alpha} \mathscr{P}(V_\beta)}} \leqslant$ $\leqslant \sum_{\beta<\alpha} \overline{\overline{\mathscr{P}(V_\beta)}} = \sup(\overline{\overline{\alpha}}, \sup_{\beta<\alpha} 2^{\overline{\overline{V_\beta}}})$, by the corollary on p. 33. But if $\beta < \alpha$, then $\overline{\overline{V_\beta}} < \lambda$ by the definition of α, and so $2^{\overline{\overline{V_\beta}}} < \lambda$, since λ is inaccessible. For the same reason, the supremum of the family of cardinals $(2^{\overline{\overline{V_\beta}}})_{\beta<\alpha}$, being indexed by $\alpha < \lambda$, must be less than λ. Thus V_α is not greater in cardinality than the supremum of two cardinals each strictly $< \lambda$; so $\overline{\overline{V_\alpha}} < \lambda$, which contradicts our assumption about α.

So $\overline{\overline{V_\alpha}} < \lambda$ for all $\alpha < \lambda$, which means that $\overline{\overline{V_\lambda}} = \sup_{\alpha<\lambda} \overline{\overline{V_\alpha}} \leqslant \lambda$, and so $\overline{\overline{V_\lambda}} = \lambda$.

If $a \in V_\lambda$ then $a \in V_\alpha$, so $a \subset V_\alpha$ for some $\alpha < \lambda$, so $\overline{\overline{a}} \leqslant \overline{\overline{V_\alpha}}$; thus $\overline{\overline{a}} < \lambda$. Conversely, suppose $a \subset V_\lambda$ and $\overline{\overline{a}} < \lambda$. The map $x \to \overline{\overline{\mathrm{rn}(x)}}$ defined on a is a family of cardinals each less than λ, indexed by a set a whose cardinal is also less than λ. Thus the least upper bound of the family is some cardinal $\aleph_\rho < \lambda$.

So for every $x \in a$, $\overline{\overline{\mathrm{rn}(x)}} \leqslant \aleph_\rho$, and therefore $\mathrm{rn}(x) < \aleph_{\rho+1} = \gamma$. Thus $a \subset V_\gamma$. But $\gamma = \aleph_{\rho+1} \leqslant 2^{\aleph_\rho} < \lambda$. Consequently $a \in V_\lambda$. ▮

If λ is inaccessible then in V_λ hold all the axioms of ZF, together with AF and AC.

It is easy to check that the axioms of extensionality, union, power-set, foundation, and infinity are satisfied.

For the axiom of choice, let $a \in V_\lambda$ be a family of disjoint non-empty sets. Then (since AC is here imagined to hold in $\mathscr{U}$) there is a set $b \subset \bigcup a$ whose intersection with each $x \in a$ is a singleton. Since $a \subset V_\alpha$ for some $\alpha < \lambda$, $b \subset V_\alpha$ too, so $b \in V_{\alpha+1}$, so $b \in V_\lambda$.

For the replacement scheme we take $a \in V_\lambda$, and pick any formula $R(x, y)$, with parameters in V_λ only, that within V_λ defines a functional relation. Within $\mathscr{U}$ the functional relation is defined by the formula $x \in V_\lambda \wedge y \in V_\lambda \wedge R^{V_\lambda}(x, y)$. As its domain is a set, it corresponds to some map f, and the range of f is a set $b \subset V_\lambda$. Since trivially $\bar{b} \leqslant \bar{a} < \lambda$, it follows that $b \in V_\lambda$.

The axiom of accessibility *ACC* denies the existence of inaccessible cardinals: *All cardinals are accessible.*

We can prove the relative consistency of *ACC* (relative, that is, to *ZF*+*AC*) as follows. Take a universe $\mathscr{U}$ where *AC* holds. Either *ACC* does too – in which case nothing needs to be proved – or there is an inaccessible cardinal in $\mathscr{U}$, the first such being π say. Then, by the above, V_π satisfies *ZF*+*AF*+*AC*. We shall see that it satisfies *ACC* too.

The ordinals of V_π *are the ordinals* $<\pi$. For if α is an ordinal of V_π, it is a transitive element of V_π linearly ordered by $\in$. But then it is an ordinal in $\mathscr{U}$; since it must be in V_π it is an ordinal less than π.

The cardinals of V_π *are the cardinals* $<\pi$. If α is a cardinal $<\pi$, we have $\alpha \in V_\pi$; and within V_π there is no one-one map of α on to a smaller ordinal, since there is not one even in $\mathscr{U}$ itself. Thus α is a cardinal of V_π.

Conversely, if $\alpha<\pi$ is not a cardinal we can map it by some bijection f on to a smaller ordinal γ. So $f \subset \alpha\times\gamma$, so $f \subset \alpha\times\alpha$; thus $f \in V_{\alpha+3}$. Consequently $f \in V_\pi$ and there is in V_π a bijection from α on to γ, which proves that α is not a cardinal of V_π.

So let α be a cardinal $<\pi$; by the definition of π it is accessible in $\mathscr{U}$, and so one of (1), (2), (3) above must fail for α. Therefore one of the following holds.

(1′) $\alpha \leqslant \omega$; here α must be accessible in V_π.

(2′) There is a cardinal $\beta<\alpha$ for which $2^\beta \geqslant \alpha$; then $\beta \in V_\pi$, $2^\beta \in V_\pi$, and so α is accessible in V_π.

(3′) $\alpha \leqslant \bigcup_{i\in I} \beta_i$, where I is a cardinal less than α and every β_i is likewise a cardinal $<\alpha$. The map $i \to \beta_i$ is a subset of $I\times\alpha$ and therefore of $\alpha\times\alpha$; so it belongs to V_π. Thus α is accessible in V_π.

This suffices to establish that the axiom of accessibility holds in V_π.

CHAPTER IV

THE REFLECTION PRINCIPLE

Let X be a class and $E(x_0, \ldots, x_{k-1})$ a formula with all parameters confined to X. We shall say that X *mirrors* the formula E (or is a *mirror* for E) if, for arbitrary $a_0, \ldots, a_{k-1}$ in X, E holds if and only if E^X also holds. That is, 'the class X mirrors the formula E' is none other than

$$\forall x_0 \ldots \forall x_{k-1} [X(x_0) \wedge \ldots \wedge X(x_{k-1}) \\ \rightarrow (E(x_0, \ldots, x_{k-1}) \leftrightarrow E^X(x_0, \ldots, x_{k-1}))].$$

Note that if E is quantifier-free then E^X is E itself, and so any class whatever is a mirror.

A formula E is said to be *prenex* (or *in prenex normal form*) if it is $N_0 N_1 \ldots N_{r-1} E'$, where each N_i either is $\neg$ or is $\exists$ followed by some variable x, and E' is free of quantifiers. For prenex formulas E, F we say that F is a *truncation* of E if it is obtained from the latter by removing some of the $\neg$'s and $\exists x$'s at the beginning. So if E is $N_0 N_1 \ldots N_{r-1} E'$, for some $p > 0$ the formula F is of the form $N_p N_{p+1} \ldots N_{r-1} E'$.

It is well known, and will not be proved here, that every formula A is equivalent to a prenex formula E containing exactly the same parameters and exactly the same free variables (see [4] or [5]).

LEMMA: *Let E be a prenex formula without parameters, and let $(X_n)_{n \in \omega}$ be an increasing sequence of sets whose limit (that is, $\bigcup_{n \in \omega} X_n$) is X. If each X_n is a mirror for E and all its truncations, then X too is a mirror for E and all truncations.*

PROOF: We prove this lemma by informal (metamathematical) induction on the number of $\neg$'s and $\exists x$'s at the front of E. When E is quantifier-free there is nothing to prove since either E has no truncations or all its truncations are quantifier-free as well; in either case every set is a mirror.

If E is $\neg F$ suppose that each X_n mirrors E and all its truncations; then it obviously mirrors F and all *its* truncations, and so by the induction hypothesis X mirrors F and all truncations. Consequently,

$$\forall x_0 \dots \forall x_{k-1} [X(x_0) \wedge \dots \wedge X(x_{k-1}) \\ \rightarrow (F(x_0, \dots, x_{k-1}) \leftrightarrow F^X(x_0, \dots, x_{k-1}))],$$

and since E is $\neg F$, E^X is $\neg F^X$, and so X mirrors E.

Likewise, if E is $\exists x F(x, x_0, \dots, x_{k-1})$ and every X_n mirrors E and its truncations, then every X_n mirrors F and its truncations; so (induction hypothesis) X mirrors F and its truncations. It remains to show that X is a mirror for E.

Pick $a_0, \dots, a_{k-1}$ in X. If $E^X(a_0, \dots, a_{k-1})$ is true we have $\exists x(x \in X \wedge \wedge F^X(x, a_0, \dots, a_{k-1}))$, and so there is an $a \in X$ for which $F^X(a, a_0, \dots, a_{k-1})$. As X mirrors F, this gives $F(a, a_0, \dots, a_{k-1})$ and so $\exists x F(x, a_0, \dots, a_{k-1})$. But this is $E(a_0, \dots, a_{k-1})$, which is thereby established.

For the converse suppose that $E(a_0, \dots, a_{k-1})$. By taking n sufficiently big we can ensure that each of $a_0, \dots, a_{k-1}$ is in X_n (since they are in X). As X_n is a mirror for E we get $E^{X_n}(a_0, \dots, a_{k-1})$, or

$$\exists x [x \in X_n \wedge F^{X_n}(x, a_0, \dots, a_{k-1})].$$

Thus there is an element a of X_n for which $F^{X_n}(a, a_0, \dots, a_{k-1})$. Now X_n mirrors F, so it follows from this that $F(a, a_0, \dots, a_{k-1})$. And X also mirrors F, so F^X $(a, a_0, \dots, a_{k-1})$ too. But this entails that $\exists x(x \in X \wedge \wedge F^X$ $(x, a_0, \dots, a_{k-1}))$ or, what is the same thing, $E^X(a_0, \dots, a_{k-1})$. And this proves the lemma. ▮

Note that this lemma is in effect a lemma scheme, since for each formula E it yields the proof of a true formula depending on E.

The *reflection principle* is another theorem scheme derivable (if AF is assumed) within the theory ZF. According to the principle, *for every parameter-free formula* $E(x_0, \dots, x_{k-1})$ *there exists an arbitrarily large limit ordinal* β *for which* V_β *mirrors* E. More formally, we state it as follows.

THEOREM: *For every parameter-free formula* $E(x_0, \dots, x_{k-1})$ *we have*

$$\forall \alpha \exists \beta > \alpha \forall x_0 \dots \forall x_{k-1} [\beta \text{ is a limit ordinal} \\ \wedge (x_0 \in V_\beta \wedge \dots \wedge x_{k-1} \in V_\beta \\ \rightarrow (E(x_0, \dots, x_{k-1}) \leftrightarrow E^{V_\beta}(x_0, \dots, x_{k-1})))].$$

PROOF: We suppose E to be in prenex normal form, and prove by informal induction on its length that for each ordinal α there is a greater limit ordinal β such that V_β mirrors E and all truncations.

If E is quantifier-free we take for β the first limit ordinal $>\alpha$, since every set – and therefore V_β – is a mirror for E.

If E is $\neg F$ we can find a limit $\beta > \alpha$ such that V_β mirrors F and all truncations of F. So for $a_0, \ldots, a_{k-1}$ in V_β we get

$$F(a_0, \ldots, a_{k-1}) \leftrightarrow F^{V_\beta}(a_0, \ldots, a_{k-1}),$$

and so

$$E(a_0, \ldots, a_{k-1}) \leftrightarrow E^{V_\beta}(a_0, \ldots, a_{k-1});$$

thus V_β also mirrors E.

So suppose E is $\exists x F(x, x_0, \ldots, x_{k-1})$; according to the induction hypothesis, whatever ordinal α is there is a larger limit β such that V_β mirrors F and all its truncations.

We define the functional relation $y = \Phi(x_0, \ldots, x_{k-1})$ of k arguments to mean that 'y is the set of all x of minimal rank for which $F(x, x_0, \ldots, x_{k-1})$ holds, if there are such x, and the empty set otherwise'. It is clear from the definition that

$$\exists x F(x, x_0, \ldots, x_{k-1}) \leftrightarrow \exists x \left[x \in \Phi(x_0, \ldots, x_{k-1}) \wedge F(x, x_0, \ldots, x_{k-1})\right].$$

We now proceed to define an ω-sequence of ordinals, $(\beta_n)_{n \in \omega}$, as follows.

β_0 is the first ordinal $>\alpha$ such that V_{β_0} is a mirror for F and all its truncations.

When β_i is defined for each $i \leqslant 2n$ we put β_{2n+1} equal to the first ordinal $>\beta_{2n}$ such that

$$\Phi(a_0, \ldots, a_{k-1}) \subset V_{\beta_{2n+1}}$$

for every k-tuple $a_0, \ldots, a_{k-1}$ drawn from $V_{\beta_{2n}}$. (Since

$$\bigcup \{\Phi(a_0, \ldots, a_{k-1}) \mid \langle a_0, \ldots, a_{k-1}\rangle \in V^k_{\beta_{2n}}\}$$

is a set, it is included in some V_γ, and so β_{2n+1} is well defined.)

β_{2n+2}, on the other hand, is put equal to the first ordinal $\gamma > \beta_{2n+1}$ for which we get a mirror V_γ for F and all truncations.

The sequence β_n is strictly increasing, as is plain enough. So if $\beta = \sup_{n \in \omega} \beta_n$, β is a limit ordinal and $V_\beta = \bigcup_{n \in \omega} V_{\beta_n} = \bigcup_{n \in \omega} V_{\beta_{2n}}$. But, by stipulation, each $V_{\beta_{2n}}$ mirrors F and all truncations; and therefore, by the lemma, V_β is also a mirror for F.

It remains to prove that V_β is a mirror for E. Take, then, any $a_0, \dots, a_{k-1} \in V_\beta$, and fix n large enough to fit all these objects into $V_{\beta_{2n}}$.

If $E(a_0, \dots, a_{k-1})$ is true we have $\exists x F(x, a_0, \dots, a_{k-1})$ and so $\exists x(x \in \Phi(a_0, \dots, a_{k-1}) \wedge F(x, a_0, \dots, a_{k-1}))$. This gives an $a \in \Phi(a_0, \dots, a_{k-1})$ for which $F(a, a_0, \dots, a_{k-1})$. As each of $a_0, \dots, a_{k-1}$ is itself in $V_{\beta_{2n}}$, a qualifies for $V_{\beta_{2n+1}}$, so $a \in V_\beta$. V_β, however, mirrors F, so $F^{V_\beta}(a, a_0, \dots, a_{k-1})$, so $\exists x(x \in V_\beta \wedge F^{V_\beta}(x, a_0, \dots, a_{k-1}))$, so $E^{V_\beta}(a_0, \dots, a_{k-1})$ as required.

Conversely, given the latter fact there must exist an $a \in V_\beta$ verifying $F^{V_\beta}(a, a_0, \dots, a_{k-1})$. Since V_β mirrors F, we have then $F(a, a_0, \dots, a_{k-1})$ for this a, and so $E(a_0, \dots, a_{k-1})$, which concludes the proof. ∎

To see the significance of this theorem in a special case, note first that if E is a closed formula, a sentence, then to say that V_β is a mirror for it is just to say that it is true in the universe iff it is true in V_β. Thus if E is a sentence which is true in the universe, there are arbitrary large limit ordinals β such that E is true in V_β.

In later parts of the book we shall make use of a somewhat more general principle of reflection. *Consider a functional relation $y = W_\alpha$ with domain On, increasing $(\alpha \leqslant \beta \rightarrow W_\alpha \subset W_\beta)$ and continuous (for limit α, $W_\alpha = \bigcup_{\beta<\alpha} W_\beta$). Let W be the union (in the intuitive sense) of all the W_α – that is, the class defined by $\exists \alpha[On(\alpha) \wedge x \in W_\alpha]$. Then, provided $E(x_0, \dots, x_{k-1})$ has no parameters, there is, for every ordinal α, a greater limit ordinal β such that*

$$\forall x_0 \dots \forall x_{k-1}[x_0 \in W_\beta \wedge \dots \wedge x_{k-1} \in W_\beta \rightarrow (E^W(x_0, \dots, x_{k-1}) \leftrightarrow E^{W_\beta}(x_0, \dots, x_{k-1}))].$$

No real modification of the proof is required to establish this form of the principle; for, apart from its continuity, the only fact about the functional relation $y = V_\alpha$ used above was that it was increasing. In this more general case, the expression 'the set X mirrors the formula E' can be taken as shorthand for the assertion that every member of X is in the class W and

$$\forall x_0 \dots \forall x_{k-1}[x_0 \in X \wedge \dots \wedge x_{k-1} \in X \rightarrow (E^W(x_0, \dots, x_{k-1}) \leftrightarrow E^X(x_0, \dots, x_{k-1}))].$$

It should be observed that the axiom of foundation is no longer needed for the more general form of the reflection principle; its only use previously was in guaranteeing that every set belonged to some V_α.

COMPARISON OF THE THEORIES OF ZERMELO AND ZERMELO/FRAENKEL

Let α be a limit ordinal $>\omega$ (α is then not less than $\omega+\omega$). There is no difficulty in showing that in the set V_α (enriched with the membership relation) all the following axioms hold.

(1) Axiom of Extensionality.

(2) Pairing Axiom: $\forall x \forall y \exists z \forall t [t \in z \leftrightarrow t = x \vee t = y]$.

(3) Union Axiom.

(4) Power-set Axiom.

(5) Axiom of Infinity: $\exists x [\emptyset \in x \wedge \forall y (y \in x \rightarrow y \cup \{y\} \in x)]$.

(6) Scheme of Comprehension: for every formula $A(x, x_0, \ldots, x_{k-1})$ without parameters, the axiom

$$\forall x_0 \ldots \forall x_{k-1} \forall x \exists y \forall z [z \in y \leftrightarrow [z \in x \wedge A(z, x_0, \ldots, x_{k-1})]].$$

(1) and (3) hold because V_α is transitive; (2) and (4) because α is a limit ordinal – for if $x, y \in V_\alpha$, we get $x, y \in V_\beta$ for some $\beta<\alpha$, so $\{x, y\} \in V_{\beta+1}$, and $\mathscr{P}(x) \in V_{\beta+1}$, where $\beta+1<\alpha$; and (5) holds because $\omega \in V_\alpha$ – we selected α above ω.

Lastly, to show that all the comprehension axioms (6) hold, suppose $a, a_0, \ldots, a_{k-1} \in V_\alpha$; they then belong to some V_β for $\beta<\alpha$, so any subset of a, for example the set $b=\{z \in a \mid A^{V_\alpha}(z, a_0, \ldots, a_{k-1})\}$ is in $V_{\beta+1}$, and so in V_α as required.

We designate by Z the theory (Zermelo set theory) axiomatized by (1), (2), (3), (4), (5), (6).

It is not hard to see that, provided Z is consistent, it is strictly weaker than ZF. For we can find a sentence – for example 'Every well-ordered system is isomorphic to an ordinal' – which is derivable in ZF but not in Z.

For suppose given some universe $\mathscr{U}$; in $V_{\omega+\omega}$ ($\omega+\omega$ is the first limit ordinal $>\omega$), as we have just seen, all the axioms of Z hold. The ordinals of $V_{\omega+\omega}$ – the sets, that is, that satisfy the relativization of $On(x)$ to $V_{\omega+\omega}$ – are simply the ordinals $<\omega+\omega$. But $\mathscr{P}(\omega\times\omega) \in V_{\omega+\omega}$, so every well-ordering of ω, being a subset of $\omega \times \omega$, is in $V_{\omega+\omega}$. Among these well-orderings is one whose ordinal is $\omega+\omega$; but this ordinal is not an ordinal of $V_{\omega+\omega}$, so the well-ordering in question is isomorphic to no ordinal. Thus $V_{\omega+\omega}$ fails to satisfy the sentence mentioned, which therefore cannot be derived in Z.

Note that the foundation axiom also holds in V_α (α being a limit ordinal beyond ω). We now intend to prove that no consistent extension of $ZF+AF$ can be obtained by adding finitely many new axioms to $Z+AF$; put another way, this means that *whatever sentence, consistent with* $Z+AF$, A *may be, we can find a formula* B *which is a consequence of* $ZF+AF+A$, *but not of* $Z+AF+A$.

The formula B is 'there exists a limit ordinal $\beta>\omega$ such that A^{V_β}'. This is certainly a consequence of $ZF+AF+A$, since A is a consequence of that theory. We need only apply the reflection principle with ω for α.

Now if α is a limit ordinal, it is easily checked that V_α is a mirror for all the following formulas (whose free variables are just ξ, x, y).

(1) $On(\xi)$, when it is written out as
$$\forall x(x \in \xi \rightarrow x \subset \xi) \wedge \forall x \forall y[x \in \xi \wedge y \in \xi \rightarrow x \in y \vee x=y \vee y \in x].$$
(2) 'ξ is a limit ordinal', that is,
$$On(\xi) \wedge \xi \neq 0 \wedge \forall x[\xi \neq x \cup \{x\}].$$
(3) $y=\mathscr{P}(x)$, that is $\forall z[z \in y \leftrightarrow z \subset x]$.

(4) $y=\bigcup x$, that is, $\forall u[u \in y \leftrightarrow \exists z(z \in x \wedge u \in z)]$.

(5) $x \in V_\xi$, that is, $On(\xi) \wedge \exists f[f$ is a map with domain $\xi+1$ such that $\forall \eta \in \xi+1\,(f(\eta)=\bigcup_{\zeta \in \eta} \mathscr{P}(f(\zeta))) \wedge x \in f(\xi)]$.

Taking these as proved we can proceed to the following lemma.

LEMMA: *Let* $E(x_0, \ldots, x_{k-1})$ *be a parameter-free formula, and* α *a limit ordinal. Then* V_α *mirrors* $E^{V_\xi}(x_0, \ldots, x_{k-1})$ *(whose free variables are* ξ, $x_0, \ldots, x_{k-1}$*).*

PROOF: We use metamathematical induction on the length of E, again assumed to be in prenex normal form. If E has no quantifiers there is nothing to prove, and if E is $\neg F$ then E^{V_ξ} is $\neg F^{V_\xi}$; so V_α mirrors E^{V_ξ} iff it mirrors F^{V_ξ} (which it does, by the induction hypothesis).

So suppose that $E(x_0, \ldots, x_{k-1})$ is $\exists x F(x, x_0, \ldots, x_{k-1})$, F itself being a formula for which the lemma holds. Then E^{V_ξ} is the formula $\exists x [x \in V_\xi \wedge \wedge F^{V_\xi}(x, x_0, \ldots, x_{k-1})]$ and so $[E^{V_\xi}(x_0, \ldots, x_{k-1})]^{V_\alpha}$ is $\exists x [x \in V_\alpha \wedge \wedge (x \in V_\xi)^{V_\alpha} \wedge (F^{V_\xi}(x, x_0, \ldots, x_{k-1}))^{V_\alpha}]$.

We have noted above (in (5)) that V_α mirrors the formula of two variables $x \in V_\xi$. As a consequence, if $x, \xi \in V_\alpha$, we get $(x \in V_\xi)^{V_\alpha} \leftrightarrow x \in V_\xi$; furthermore, since ξ is in V_α it is less than α, so $x \in V_\alpha \wedge x \in V_\xi \leftrightarrow x \in V_\xi$. And again, since V_α mirrors F^{V_ξ} (induction hypothesis), we have for $x, x_0, \ldots, x_{k-1} \in V_\alpha$ that

$$(F^{V_\xi}(x, x_0, \ldots, x_{k-1}))^{V_\alpha} \leftrightarrow F^{V_\xi}(x, x_0, \ldots, x_{k-1}).$$

Gathering together all these remarks, we can conclude that for ξ, $x_0, \ldots, x_{k-1} \in V_\alpha$ we have

$$[E^{V_\xi}(x_0, \ldots, x_{k-1})]^{V_\alpha} \leftrightarrow \exists x [x \in V_\xi \wedge F^{V_\xi}(x, x_0, \ldots, x_{k-1})];$$

and this latter being $E^{V_\xi}(x_0, \ldots, x_{k-1})$, the result follows. ∎

So let α be the first limit ordinal greater than ω for which V_α mirrors the sentence A; since A is true in $\mathscr{U}$, α is the least limit ordinal $> \omega$ such that A^{V_α}.

By the lemma V_α mirrors the formula A^{V_ξ} (of one free variable only, ξ). B is just $\exists \xi$ [ξ is a limit ordinal $> \omega \wedge A^{V_\xi}$], so B^{V_α} is $\exists \xi$ [$\xi \in V_\alpha \wedge$ (ξ is a limit ordinal $> \omega)^{V_\alpha} \wedge (A^{V_\xi})^{V_\alpha}$]. Since V_α mirrors both A^{V_ξ} and 'ξ is a limit ordinal $> \omega$', B^{V_α} is equivalent to $\exists \xi$[$\xi \in V_\alpha \wedge \xi$ is a limit ordinal $> \omega \wedge A^{V_\xi}$]. However, as α was chosen to be the *least* limit above ω for which A^{V_α}, we cannot have, for a limit ξ, both A^{V_ξ} and $\xi \in V_\alpha$; and, as we can see, B^{V_α} requires both to be true. Thus B^{V_α} is false. This proves that, though V_α satisfies all the axioms of Z, AF, and A (by the choice of α), the formula B does not hold in it. So B cannot be a consequence of $Z + AF + A$.

As a corollary to the above result we may note that *if ZF is consistent, it is not finitely axiomatizable*. For the assumption of consistency allows us to assume $ZF + AF$ consistent; so if ZF were equivalent to a single sentence A, $ZF + AF$ would be equivalent to $Z + AF + A$.

Note that even if A is an arithmetical formula (by this we mean one relativized to V_ω), B is not. Gödel's second incompleteness theorem provides us with another proof of the above result, but one which associates with every A consistent with $ZF+AF$ an *arithmetical* B' following from $ZF+AF+A$, but not from $Z+AF+A$.

The following, however, remains an open problem: if ZF is consistent, does there exist a sentence A, consistent with Z, such that $Z+A$ is equivalent to $ZF+A$? From what we have just shown it is clear that such a sentence A would have to contradict the axiom of foundation; for otherwise $Z+AF+A$ would be consistent and equivalent to $ZF+AF+A$.

CHAPTER V

THE SET OF EXPRESSIONS

In the first two chapters we constructed, within each universe $\mathscr{U}$, a sort of replica for several of the fundamental ideas of mathematics; the idea of a mapping, for example, or that of a natural number. And we agreed thenceforth to use these words in the senses we had given them in $\mathscr{U}$, and not at all in their everyday senses.

We are now going to carry out the same 'reconstruction' for the idea of a *formula of set theory*. In this case, however, we shall continue to use the word 'formula' in its intuitive sense only, as we have used it up to now, and will take advantage of the word '*expression*' in order to refer to the kind of objects that we shall define in the universe.

Choose from $\mathscr{U}$ any five distinct sets, to be denoted by $\mathbf{v}$, $\sim$, $\exists$, ε, and $=$; these sets might be, for instance, the sets 0, 1, 2, 3, 4. Choose also a denumerable set $\mathscr{V}$ whose elements, called *variables*, differ from all five sets chosen above – $\mathscr{V}$ might consist of the natural numbers $\geqslant 5$. If x is a variable we shall write the ordered pair $\langle \exists, x\rangle$ as $\exists x$.

We define by recursion a map $n \to \mathscr{E}_n$ with domain ω: $\mathscr{E}_0$ is the set of ordered triples $\langle \varepsilon, x, y\rangle$ and $\langle =, x, y\rangle$ where x, y are in $\mathscr{V}$; more formally,

$$\mathscr{E}_0 = [\{\varepsilon\} \times \mathscr{V}^2] \cup [\{=\} \times \mathscr{V}^2].$$

The elements of $\mathscr{E}_0$ are called *atomic expressions*. Further,

$$\mathscr{E}_{n+1} = \mathscr{E}_n \cup [\{\sim\} \times \mathscr{E}_n] \cup [\{\mathbf{v}\} \times \mathscr{E}_n^2] \cup [(\{\exists\} \times \mathscr{V}) \times \mathscr{E}_n];$$

it consists therefore of everything in $\mathscr{E}_n$, together with ordered pairs and triples of the form $\langle \sim, \Phi\rangle$, $\langle \mathbf{v}, \Phi, \Psi\rangle$, $\langle \exists x, \Phi\rangle$ for $\Phi, \Psi \in \mathscr{E}_n$ and $x \in \mathscr{V}$.

Writing $\mathscr{E}$ for $\bigcup_{n\in\omega} \mathscr{E}_n$ we call $\mathscr{E}$ the *set of all expressions.*

It is easy to check by induction on n that, in virtue of our choices for $\mathbf{v}$, $\sim$, $\exists$, ε, $=$, and $\mathscr{V}$, every $\mathscr{E}_n \subset V_\omega$, and so $\mathscr{E} \subset V_\omega$; every expression is a hereditarily finite set.

Given an expression Φ we shall call the first n for which $\Phi \in \mathscr{E}_n$ the *depth* of Φ.

LEMMA: *For each expression Φ one and only one of the following possibilities is realized:*

Φ is atomic;
Φ is of the form $\langle \sim, \Psi \rangle$;
Φ is of the form $\langle \vee, \Psi, \Psi' \rangle$;
Φ is of the form $\langle \exists x, \Psi \rangle$.

Furthermore, Ψ and Ψ' are expressions determined by Φ, and strictly lower in depth than Φ itself.

PROOF: Bearing in mind that the ordered triple $\langle a, b, c \rangle$ is defined as the ordered pair $\langle a, \langle b, c \rangle\rangle$, it is clear that each expression is an ordered pair. Also clear is the fact that the first element of such a pair is bound to be ε, $=$, $\sim$, $\vee$, or $\exists x$ (where $x \in \mathscr{V}$); and these objects are distinct from one another, as that is the way we chose them.

If the first element of a pair is ε or $=$ we must have $\Phi \in \mathscr{E}_0$; if it is $\sim$, the second element of the pair is an expression Ψ. Writing $n+1$ for the depth of Φ we have $\langle \sim, \Psi \rangle \in \mathscr{E}_{n+1}$ and $\langle \sim, \Psi \rangle \notin \mathscr{E}_n$; so by the definition of $\mathscr{E}_{n+1}$ we get $\Psi \in \mathscr{E}_n$; and therefore the depth of Ψ is less than $n+1$. On the other hand it is obvious that Ψ is determined uniquely by Φ. Exactly the same goes if the first element of the pair is $\exists x$. If, however, it is $\vee$, then the second element takes the form $\langle \Psi, \Psi' \rangle$, Ψ and Ψ' being determined by Φ, obviously enough. From $\langle \vee, \Psi, \Psi' \rangle \in \mathscr{E}_{n+1}$ and $\langle \vee, \Psi, \Psi' \rangle \notin \mathscr{E}_n$ it again follows at once that $\Psi, \Psi' \in \mathscr{E}_n$, and therefore Ψ and Ψ' have depth less than $n+1$. ▮

In the sequel the expressions $\langle \varepsilon, x, y \rangle$, $\langle =, x, y \rangle$, $\langle \sim, \Phi \rangle$, $\langle \vee, \Phi, \Psi \rangle$, and $\langle \exists x, \Phi \rangle$ will be shortened to $x \,\varepsilon\, y$, $x=y$, $\sim\Phi$, $(\Phi) \vee (\Psi)$, and $\exists x\, \Phi$ respectively; the sets ε, $=$, $\sim$, $\vee$ will be called the *membership sign*, the *equals sign*, the *negation sign*, and the *disjunction sign* respectively; and $\exists$ is the *existential quantifier*.

The expressions $(\sim\Phi) \vee (\Psi)$, $\sim((\sim\Phi) \vee (\sim\Psi))$, $\sim\exists x(\sim\Phi)$ will be still further shortened to $(\Phi) \Rightarrow (\Psi)$, $(\Phi) \wedge (\Psi)$, $\forall x(\Phi)$ respectively; and the expression $(\Phi \Rightarrow \Psi) \wedge (\Psi \Rightarrow \Phi)$ will be written $(\Phi) \Leftrightarrow (\Psi)$.

We now define recursively on the depth of an expression a map w from $\mathscr{E}$ into the set of all finite subsets of $\mathscr{V}$; $w(\Phi)$ will be called the *set* *of free variables of the expression Φ.*

DEFINITION:

If $\Phi = x \, \varepsilon \, y$ or $\Phi = x = y$ then $w(\Phi) = \{x, y\}$;
$w(\sim \Phi) = w(\Phi)$;
$w(\Phi \vee \Psi) = w(\Phi) \cup w(\Psi)$;
$w(\exists x\Phi) = w(\Phi) \setminus \{x\}$.

An expression Φ is called *closed* if $w(\Phi) = \emptyset$.

It should be noted that each parameter-free formula A corresponds in an obvious way to an expression, an expression we shall denote by $\ulcorner A \urcorner$. But the converse is false. Were there to exist in $\mathscr{U}$ a natural number n with, intuitively speaking, infinitely many elements, then there would be an expression with n as its depth, but no formula could be expected to correspond to it. In fact, one can readily see that an expression corresponds to a formula if and only if its depth is a natural number with only a finite number of elements (in the intuitive sense of the words).

We can also define (recursively on the depth of the expression Φ) a binary functional relation $Y = \mathrm{Val}(\Phi, X)$ whose domain is 'Φ is an expression and X is a set'. Y is called the *value of the expression Φ in the set X*, and it is a subset of ${}^{w(\Phi)}X$.

DEFINITION:

(1) If Φ is $x \, \varepsilon \, y$ (respectively, $x = y$) then
$\mathrm{Val}(\Phi, X) = \{\delta \in {}^{\{x, y\}}X \mid \delta(x) \in \delta(y)\}$
(respectively $\{\delta \in {}^{\{x, y\}}X \mid \delta(x) = \delta(y)\}$);

(2) If $\Phi = \sim \Psi$ then $\mathrm{Val}(\Phi, X) = {}^{w(\Phi)}(X) \setminus \mathrm{Val}(\Psi, X)$;

(3) If $\Phi = \Psi \vee \Psi'$ then $\mathrm{Val}(\Phi, X)$
$= \{\delta \in {}^{w(\Phi)}X \mid \delta \upharpoonright w(\Psi) \in \mathrm{Val}(\Psi, X)$
$\vee \ \delta \upharpoonright w(\Psi') \in \mathrm{Val}(\Psi', X)\}$;

(4) If $\Phi = \exists x \Psi$ then $\mathrm{Val}(\Phi, X)$
$= \{\delta \in {}^{w(\Phi)}X \mid \exists \delta' \supset \delta(\delta' \in {}^{w(\Psi)}X \wedge \delta' \in \mathrm{Val}(\Psi, X))\}$
(remember that in this case $w(\Phi) = w(\Psi) \setminus \{x\}$).

If Φ is the expression corresponding to the formula $A(x_0, \ldots, x_{k-1})$ it is simple to see that $\mathrm{Val}(\Phi, X)$ is the set of maps δ from $\{x_0, \ldots, x_{k-1}\}$ into

X for which $A^X(\delta(x_0),\ldots,\delta(x_{k-1}))$ holds; this is easily proved by informal metamathematical induction on the length of the formula A.

We shall write an expression Φ whose free variables form a finite set $\{x_0,\ldots,x_{k-1}\}$ (indexed by a natural number k) as $\Phi(x_0,\ldots,x_{k-1})$; such an expression will, for brevity, be called *k-adic*.

An *expression with parameters* Φ_0 is by definition an ordered pair $\langle\Phi,\eta\rangle$ where Φ is an expression and η a map defined on some subset of $w(\Phi)$ (perhaps $w(\Phi)$ itself). If the free variables of Φ are $x_0,\ldots,x_{k-1}$, $y_0,\ldots,y_{l-1}$ (and they are all different), and the domain of η is the subset $\{x_0,\ldots,x_{k-1}\}$ of $w(\Phi)$, and the value of η at each x_i (for $i<k$) is a_i, then the expression Φ_0 with parameters is written $\Phi(a_0,\ldots,a_{k-1},y_0,\ldots,y_{l-1})$. The set $\{y_0,\ldots,y_{l-1}\}$ – that is, $w(\Phi)\setminus dom(\eta)$ – is, by definition, the set of free variables of the expression Φ_0 with parameters, and is denoted by $w(\Phi_0)$.

Suppose now that X is a set and $\Phi_0=\langle\Phi,\eta\rangle$ an expression all of whose parameters are taken from X. Such an expression will sometimes be called an X-expression. We extend the earlier definition of Val so that $\mathrm{Val}(\Phi_0,X)$ becomes the set of all maps $\delta\in{}^{w(\Phi_0)}X$ such that $\delta\cup\eta\in\mathrm{Val}(\Phi,X)$ – here $\delta\cup\eta$ is the map defined on $w(\Phi)$ as equal to δ on $w(\Phi_0)$ and to η on $w(\Phi)\setminus w(\Phi_0)$.

If $A(x_0,\ldots,x_{k-1})$ is any formula whatever with parameters, there corresponds to it in an obvious fashion an expression with parameters, $\ulcorner A\urcorner$ for short. $\mathrm{Val}(\ulcorner A\urcorner,X)$ will be properly defined if all the parameters of the formula A are elements of X; in which case $\mathrm{Val}(\ulcorner A\urcorner,X)$ is the set of all maps δ from $\{x_0,\ldots,x_{k-1}\}$ into X for which the formula $A^X(\delta(x_0),\ldots,\delta(x_{k-1}))$ holds; this again follows easily by informal induction on the length of A.

An expression Φ with parameters is said to be *closed* if $w(\Phi)=\emptyset$. If Φ is a closed X-expression – so that the set X contains all the parameters of Φ – then $\mathrm{Val}(\Phi,X)$ is a subset of ${}^{\emptyset}X=\{\emptyset\}$. Accordingly, $\mathrm{Val}(\Phi,X)=1$ or 0. If it is 1 we say that Φ is *satisfied in the set X*.

The following theorem, proved with the help of the axiom of choice, will be used in Chapter VIII.

THEOREM (LÖWENHEIM/SKOLEM): *Let X be a set, P a subset of it, and $\mathscr{A}$ the set of all closed P-expressions which are satisfied in X. Then there is a* *subset $Y \subset X$ such that every expression in $\mathscr{A}$ is satisfied in Y, $\overline{\overline{Y}} \leqslant \overline{\overline{P}} + \aleph_0$, and $P \subset Y$.*

PROOF: The axiom of choice is needed to ensure a map $\theta : \mathscr{P}(X) \setminus \{\emptyset\} \to X$ such that for every non-empty $U \subset X$, $\theta(U) \in U$.

Using the map θ we define recursively an ω-sequence P_n of subsets of X: P_0 is just P itself; and P_{n+1} is defined from P_n in the following manner.

Let $\mathscr{G}_n$ be the set of monadic P_n-expressions $\Phi(x, a_0, \ldots, a_{r-1})$ for which $\exists x \Phi(x, a_0, \ldots, a_{r-1})$ is satisfied in X. Let $U_\Phi = \{\xi \in X \mid \Phi(\xi, a_0, \ldots, a_{r-1})$ is satisfied in $X\}$. Then P_{n+1} is defined as the set of all $\theta(U_\Phi)$ for Φ in $\mathscr{G}_n$.

Thus P_{n+1} contains, for each monadic $\Phi \in \mathscr{G}_n$, at least one 'satisfier' of Φ in X, namely $\theta(U_\Phi)$.

Observe first of all that $P_{n+1} \supset P_n$; for if $a \in P_n$ then the expression $x = a$ belongs to $\mathscr{G}_n$, and, for this expression, $U_\Phi = \{a\}$; thus $\theta(U_\Phi) = a$, and $a \in P_{n+1}$.

Secondly, since the set of expressions with parameters in P_n has cardinality $\overline{\overline{P_n}} + \aleph_0$, we have $\overline{\overline{\mathscr{G}_n}} \leqslant \overline{\overline{P_n}} + \aleph_0$. But the map sending the expression Φ to $\theta(U_\Phi)$ is by definition a surjection, a map from $\mathscr{G}_n$ on to P_{n+1}. Thus $\overline{\overline{P_{n+1}}} \leqslant \overline{\overline{P_n}} + \aleph_0$. By induction, therefore, $\overline{\overline{P_n}} \leqslant \overline{\overline{P}} + \aleph_0$.

We set $Y = \bigcup_{n \in \omega} P_n$. Then $P \subset Y$; and $\overline{\overline{Y}} \leqslant \sum_{n \in \omega} \overline{\overline{P_n}} \leqslant \aleph_0 \cdot (\overline{\overline{P}} + \aleph_0) = \overline{\overline{P}} + \aleph_0$.

Now let $\Phi(a_0, \ldots, a_{k-1})$ be any closed expression whose parameters are in Y. We show by induction on its depth that Φ is satisfied in X if and only if it is satisfied in Y.

For atomic Φ the result is obvious. If Φ is $\sim \Psi$ (or $\Psi \vee \Psi'$), Φ is satisfied in Y iff Ψ is not (or one or other of Ψ, Ψ' is); the induction hypothesis then tells us that Φ is satisfied in Y iff Ψ is not satisfied in X (or either Ψ or Ψ' is satisfied in X); so that Φ is satisfied in Y iff it is satisfied in X.

If $\Phi(a_0, \ldots, a_{k-1}) = \exists x \Psi(x, a_0, \ldots, a_{k-1})$ we suppose to start with that Φ is satisfied in X. Then, for a big enough n, we will have $a_0, \ldots, a_{k-1}$ all in P_n, and therefore $\Psi(x, a_0, \ldots, a_{k-1}) \in \mathscr{G}_n$. Then

$$\theta(U_\Psi) = a,$$

say, where $a \in P_{n+1}$, and so $\Psi(a, a_0, \ldots, a_{k-1})$ is satisfied in X (by the definitions of θ and U_Ψ). The depth of $\Psi(a, a_0, \ldots, a_{k-1})$, however, is strictly less than that of Φ, and so, by the induction hypothesis, $\Psi(a, a_0,$

..., a_{k-1}) is satisfied in Y; from this it follows at once that $\exists x\Psi(x, a_0, \ldots, a_{k-1})$, which is just $\Phi(a_0, \ldots, a_{k-1})$, is satisfied in Y.

Conversely, if $\exists x\Psi(x, a_0, \ldots, a_{k-1})$ is satisfied in Y, there must be an $a \in Y$ for which $\Psi(a, a_0, \ldots, a_{k-1})$ is satisfied in Y. By the induction hypothesis $\Psi(a, a_0, \ldots, a_{k-1})$ is also satisfied in X, so $\exists x\Psi(x, a_0, \ldots, a_{k-1})$ is too.

In particular we have proved that every expression in $\mathscr{A}$ is satisfied in Y. ▮

The relativization of an expression to a set: Suppose we are presented with an expression $\Phi(x_0, \ldots, x_{k-1}, a_0, \ldots, a_{l-1})$ whose parameters $a_0, \ldots, a_{l-1}$ are all in a. We define, recursively on the depth of Φ, another expression $\Phi^a(x_0, \ldots, x_{k-1}, a_0, \ldots, a_{l-1})$ known as the *relativization of Φ to the set a.*

DEFINITION: If Φ is atomic then $\Phi^a = \Phi$;
If $\Phi = \sim\Psi$ then $\Phi^a = \sim\Psi^a$;
If $\Phi = \Psi_1 \vee \Psi_2$ then $\Phi^a = \Psi_1^a \vee \Psi_2^a$;
If $\Phi = \exists x\Psi$ then $\Phi^a = \exists x(x \,\varepsilon\, a \wedge \Psi^a)$.

Clearly enough the expressions Φ, Φ^a have the same free variables. The parameters of Φ^a, however, are $a, a_0, \ldots, a_{l-1}$.

We shall use the following theorem in Chapter VIII.

THEOREM: *Suppose that $a \in b$ and $a \subset b$. If Φ is an expression with parameters in a, then $Val(\Phi, a) = {}^{w(\Phi)}a \cap Val(\Phi^a, b)$.*

PROOF: The proof is by induction on the depth of Φ. For atomic Φ the result is obvious. For molecular Φ, suppose $\Phi = \sim\Psi$ (the proof is more or less identical, *mutatis mutandis*, for $\Phi = \Psi_1 \vee \Psi_2$). Then

$$
\begin{aligned}
\mathrm{Val}(\Phi, a) &= {}^{w(\Phi)}a \setminus \mathrm{Val}(\Psi, a)\\
&= {}^{w(\Phi)}a \setminus \mathrm{Val}(\Psi^a, b)\\
&= {}^{w(\Phi)}a \cap \mathrm{Val}(\Phi^a, b),
\end{aligned}
$$

since, by hypothesis, the theorem holds for Ψ.

Now suppose Φ to be $\exists x\Psi$. We have that

$$
\mathrm{Val}(\Phi, a) = \{\delta \in {}^{w(\Phi)}a \mid \exists \delta' \supset \delta(\delta' \in {}^{w(\Psi)}(a) \wedge \delta' \in \mathrm{Val}(\Psi, a))\}.
$$

The induction hypothesis gives $\mathrm{Val}(\Psi, a)$ equal to ${}^{w(\Psi)}a \cap \mathrm{Val}(\Psi^a, b)$, so

$$\begin{aligned}\mathrm{Val}(\Phi, a) &= \{\delta \in {}^{w(\Phi)}a \mid \exists \delta' \supset \delta(\delta' \in {}^{w(\Psi)}a \wedge \delta' \in \mathrm{Val}(\Psi^a, b))\} \\ &= \{\delta \in {}^{w(\Phi)}a \mid \delta \in \mathrm{Val}(\mathbf{\exists} x(x \,\varepsilon\, a \wedge \Psi^a), b)\} \\ &= {}^{w(\Phi)}a \cap \mathrm{Val}(\Phi^a, b). \blacksquare\end{aligned}$$

CHAPTER VI

ORDINAL DEFINABLE SETS

Relative Consistency of the Axiom of Choice

Notation: Let $\Phi(x)$ be a monadic expression whose parameters are all in some set X; the value of this expression in X is then a subset of ${}^{\{x\}}X$. By the canonical bijection of ${}^{\{x\}}X$ on to X (that which sends $\{\langle x, u\rangle\}$ to u, for every $u \in X$) there corresponds to this subset a subset of X itself, written $\mathrm{val}(\Phi, X)$, and also called, by a slight abuse of language, the *value* of Φ in X.

We now consider a universe $\mathscr{U}$ where the axiom of foundation holds, and define a class OD, the *ordinal definable sets*, by the following formula, $OD(x)$: 'There are an ordinal α and a monadic expression $\Phi(y, \alpha_0, \ldots, \alpha_{r-1})$ whose parameters are ordinals $<\alpha$ and whose value in the set V_α is $\{x\}$'.

LEMMA: *Let a be a set, and $A(x, \alpha_0, \ldots, \alpha_{k-1})$, whose parameters are ordinals, be a formula with a single free variable. If a is the only set for which A holds, then a is ordinal definable.*

PROOF: Since we have chosen a universe $\mathscr{U}$ in which AF holds, we can make use of the reflection principle to obtain an ordinal α not only greater than $\alpha_0, \ldots, \alpha_{k-1}$ but also large enough for a to be in V_α, and such that V_α mirrors the formula $A(x, \alpha_0, \ldots, \alpha_{k-1})$. Then a is the only member of V_α for which the formula $A^{V_\alpha}(x, \alpha_0, \ldots, \alpha_{k-1})$ holds, and so the value of the expression $\ulcorner A(x, \alpha_0, \ldots, \alpha_{k-1})\urcorner$ in the set V_α is $\{a\}$. Thus a is ordinal definable. ∎

CONVERSE: *Let a be ordinal definable. Then there is a formula $A(x, \delta_0)$ of one free variable and one parameter (the ordinal δ_0) which holds for a and a alone.*

PROOF: Consider the parameter-free formulas $s = J(\alpha)$ and $x = K(n)$ which establish, respectively, bijections from On on to $\sigma(On)$ (the class of finite sequences of ordinals), and from ω on to V_ω. These formulas have been discussed above on pp. 33 and 41 respectively.

With the help of J one can easily define a parameter-free functional relation $\eta = J'(\alpha)$ which is a surjection from On on to the class of functions η with a finite domain included in $\mathscr{V}$ and with values in On.

Since a is ordinal definable, there are an expression $\Phi_0(x, \alpha_0, \ldots, \alpha_{r-1})$ with parameters in On, and an ordinal γ_0, such that $\mathrm{val}(\Phi_0, V_{\gamma_0}) = \{a\}$.

We therefore consider the following quaternary parameter-free formula $E(x, n, \gamma, \beta)$: 'n is a natural number, β, γ are ordinals, the ordered pair $\langle K(n), J'(\beta)\rangle$ is an expression Φ whose parameters are ordinals less than γ, and $\mathrm{val}(\Phi, V_\gamma) = \{x\}$'.

Writing then n_0 for that natural number such that $K(n_0)$ is the parameter-free expression $\Phi_0(x, x_0, \ldots, x_{r-1})$ and β_0 for that ordinal such that $J'(\beta_0)$ is the function $\eta_0 = \{\langle x_0, \alpha_0\rangle, \ldots, \langle x_{r-1}, \alpha_{r-1}\rangle\}$, it is apparent that the formula $E(x, n_0, \gamma_0, \beta_0)$ holds for a and for nothing else; this formula, however, has three parameters, the ordinals n_0, γ_0, and β_0.

To reduce the number of parameters to one, we write δ_0 for the inverse image under J of the sequence n_0, γ_0, β_0; then the formula $A(x, \delta_0)$: 'there are a natural number n and two ordinals γ, β, such that $J(\delta_0)$ is the sequence $\{\langle 0, n\rangle, \langle 1, \gamma\rangle, \langle 2, \beta\rangle\}$ and $E(x, n, \gamma, \beta)$' has δ_0 as its sole parameter, and holds for a and a alone. ∎

A formula like $A(x, \delta_0)$ is said to be a *definition of a in terms of ordinals*, or an *ordinal definition* of a.

Note that what we have actually produced above is a parameter-free formula $A(x, y)$ of two free variables such that if a is ordinal definable there is an ordinal δ_0 for which $A(x, \delta_0)$ provides a definition of a. In other words, the following theorem: $\forall x[OD(x) \leftrightarrow \exists\delta(On(\delta) \wedge \forall u(A(u, \delta) \leftrightarrow u = x))]$.

The class OD is not necessarily a transitive class, for there may be ordinal definable sets a not all of whose elements are ordinal definable. However, we can readily isolate a subclass HOD of OD, the *hereditarily ordinal definable sets*, which *is* transitive, by the formula $HOD(x)$: 'Every element of $\mathscr{C}(\{x\})$ is ordinal definable'. (Remember that $\mathscr{C}(a)$ is the transitive closure of a, the smallest transitive set, that is, which includes a; and $\mathscr{C}(\{x\}) = \{x\} \cup \mathscr{C}(x)$ – see p. 38.)

It should be observed that HOD has been defined by a formula *without* parameters.

LEMMA: *A set a is hereditarily ordinal definable if and only if it is ordinal definable and all its elements are hereditarily ordinal definable.*

PROOF: The condition is obviously necessary. So suppose that every element of a is in HOD, and that a is in OD. We must show then that a is in HOD.

From the properties of transitive closures investigated on p. 38 we have that $\mathscr{C}(\{a\})=\{a\} \cup \mathscr{C}(a)=\{a\} \cup a \cup \bigcup_{y \in a} \mathscr{C}(y)$, and so

$$\mathscr{C}(\{a\})=\{a\} \cup \bigcup_{y \in a} (\{y\} \cup \mathscr{C}(y))$$
$$=\{a\} \cup \bigcup_{y \in a} \mathscr{C}(\{y\}).$$

By assumption every element of $\mathscr{C}(\{y\})$ is in OD, whatever element y may be of a. Since a is in OD, this shows that every element of $\mathscr{C}(\{a\})$ is in OD; and so $HOD(a)$. ∎

Our main task in this chapter will be the proof of the relative consistency of the axiom of choice. We do this by constructing from a model $\mathscr{U}_0$ of ZF another model, one for $ZF+AF+AC$. Moving off from $\mathscr{U}_0$ we first construct a model $\mathscr{U}$, as we have done before (p. 43), for $ZF+AF$. We then show that not only $ZF+AF$ but also AC holds in the class HOD constructed within $\mathscr{U}$.

Axiom of Extensionality: Obvious, since if a, b are in HOD, so are all their elements.

Union Axiom: If a is in HOD, let $b=\bigcup a$. It is clear that every element of b is in HOD. According to the above lemma, all that remains to be shown is that b is ordinal definable.

Now a is certainly ordinal definable, so is the only set for which a certain formula $A(x, \alpha)$ holds; as b is the only set for which

$$B(y, a): \ \forall z(z \in y \leftrightarrow \exists u(u \in a \wedge z \in u))$$

holds, we can combine the two into

$$\exists x[A(x, \alpha) \wedge B(y, x)]$$

which is a formula of one free variable, and one ordinal parameter. It obviously holds just for b, so by the lemma on p. 64, b is ordinal definable.

Power-set Axiom: Suppose a to be in HOD, and b the set of all hereditarily ordinal definable subsets of a. We want to show $HOD(b)$, and since we have at once $HOD(x)$ for every x in b, we need only check that $OD(b)$.

If $A(x, \alpha)$ is an ordinal definition of a, since b is the only set for which

$$B(y, a):\quad \forall z[z \in y \leftrightarrow HOD(z) \wedge z \subset a]$$

holds, it is clear that $\exists x[A(x, \alpha) \wedge B(y, x)]$ holds for b and for nothing else. Thus $OD(b)$.

Scheme of Replacement: Let a be some object of HOD, and $R(x, y, a_0, \ldots, a_{k-1})$ be a binary relation with parameters in HOD. If, interpreted in HOD, R defines a functional relation, then the formula $HOD(x) \wedge \wedge HOD(y) \wedge R^{HOD}(x, y, a_0, \ldots, a_{k-1})$ defining it in the universe $\mathscr{U}$ also defines a functional relation, $S(x, y, a_0, \ldots, a_{k-1})$ say, for brevity. Let b be the set of images under S of elements of a. Since every element of b is in HOD, we need only prove that $OD(b)$. So let $A(x, \alpha)$, $A(x_0, \alpha_0), \ldots, A(x_{k-1}, \alpha_{k-1})$ be ordinal definitions of $a, a_0, \ldots, a_{k-1}$ respectively.

Then since b is the only object for which the formula

$$\begin{aligned} &B(y, a, a_0, \ldots, a_{k-1}): \\ &\forall z[z \in y \leftrightarrow \exists t(t \in a \wedge S(t, z, a_0, \ldots, a_{k-1}))] \end{aligned}$$

holds, it enjoys the same distinction with respect to the formula

$$\begin{aligned} &\exists x \exists x_0 \ldots \exists x_{k-1}[A(x, \alpha) \wedge A(x_0, \alpha_0) \\ &\qquad \wedge \ldots \wedge A(x_{k-1}, \alpha_{k-1}) \wedge B(y, x, x_0, \ldots, x_{k-1})], \end{aligned}$$

whose only parameters are the ordinals $\alpha, \alpha_0, \ldots, \alpha_{k-1}$. Thus $OD(b)$, as required.

Axiom of Infinity: If α is any ordinal it is obviously ordinal defined by $x = \alpha$; so every ordinal is in fact in HOD. $HOD(\omega)$ in particular, and the axiom of infinity holds for ω in HOD.

Axiom of Foundation: If a non-empty set a is in HOD, and b is one of its elements bearing minimal rank, then b is in HOD and $b \cap a = \emptyset$.

Axiom of Choice: We can define, by a formula without parameters as follows, an injective functional relation $\alpha = \Theta(x)$ which associates an ordinal with each monadic expression whose parameters are in On. Starting from such an expression $\Phi(x, \alpha_0, \ldots, \alpha_{r-1})$ we go first to the parameter-free expression $\Phi(x, x_0, \ldots, x_{r-1})$, which is an element of

V_ω. Using our bijection K from ω on to V_ω, we let n be the natural number for which $K(n)=\Phi(x, x_0,\dots, x_{r-1})$. In a similar way we let β be the first ordinal for which $J'(\beta)$ is the function $\{\langle x_0, \alpha_0\rangle,\dots, \langle x_{r-1}, \alpha_{r-1}\rangle\}$ where J' is the functional relation defined on p. 64. Then $\alpha=\Theta(\Phi(x, \alpha_0,\dots, \alpha_{r-1}))$ is taken as that ordinal α associated with the pair $\langle n, \beta\rangle$ by the isomorphism between On^2 and On.

Granted Θ then, we proceed to define another functional relation $\beta=D(x)$, from OD into On, by the formula 'β is the least ordinal to represent a pair of ordinals $\langle\alpha, \gamma\rangle$ where $\gamma=\Theta(\Phi)$ for some monadic expression Φ with parameters in α and value $\{x\}$ in V_α'.

The formula $\beta=D(x)$ has no parameters; what is more, it defines an injection, for $x\neq x' \to D(x)\neq D(x')$, as is easily checked.

It is clear now that the relation $R(x, y)$ defined by the parameter-free formula $OD(x) \wedge OD(y) \wedge D(x)\leqslant D(y)$ is a well-ordering of the class OD.

So suppose a to be any hereditarily ordinal definable set. Restricting the well-ordering R to a yields the set

$$b = \{\langle x, y\rangle \in a^2 \mid D(x) \leqslant D(y)\},$$

and it is this set which we shall show to be a well-ordering of a, and itself in HOD.

Taking the latter point first, note that all of b's elements are in HOD, so we have only to prove that $OD(b)$. But since the formula

$$B(y, a):\quad \forall z[z \in y \leftrightarrow \exists u\exists v(u \in a \wedge v\in a \wedge z = \langle u, v\rangle \wedge D(u) \leqslant D(v))]$$

holds only for b, and we may assume that some ordinal definition of a, $A(x, \alpha)$, holds only for a, it is clear once more that b is ordinal defined by the formula $\exists x[A(x, \alpha) \wedge B(y, x)]$. So $HOD(b)$.

As far as $\mathscr{U}$ is concerned, b certainly well-orders a; that this is also the case in HOD is apparent from the fact that all non-empty subsets of a, those lying in HOD in particular, have a smallest element (mod. b).

Thus Zermelo's theorem holds in HOD; so AC too. This concludes the proof that AC is consistent relative to ZF.

The ordinals of HOD are the same as the ordinals of $\mathscr{U}$; and the natural numbers of $\mathscr{U}$ appear unchanged in HOD too. Note that since AF holds in *HOD* we can write $On(x)$ here as

$$\forall y(y \in x \to y \subset x) \wedge \forall z[(y \in x \wedge z \in x) \\ \to (z \in y \vee y = z \vee y \in z)].$$

Moreover, V_ω is in *HOD*. For the bijection $x = K(n)$ from ω on to V_ω provides, for every hereditarily finite set, a definition whose only parameter is a natural number. Thus everything in V_ω is in *OD*, and since V_ω is transitive, this means in *HOD* too. V_ω, being definable by the parameter-free formula 'x is the set of hereditarily finite sets', is itself in *OD*, so too in *HOD*.

Continuing on these lines, we see that if f is the map $n \to V_n$ defined on ω, then $HOD(f)$; for any element of f is a pair $\langle n, V_n \rangle$, clearly hereditarily finite, and so in *HOD*. And f can be defined by the following parameter-free formula: 'f is a map defined on ω such that $f(0) = 0$ and $f(k+1) = \mathscr{P}(f(k))$ for each $k \in \omega$'.

The upshot is that the class *HOD* is a mirror for the formula $x \in V_\omega$; in other words, that $(x \in V_\omega)^{HOD} \leftrightarrow x \in V_\omega$. For $x \in V_\omega$ can be written

$$\exists f \exists n [f \text{ is a map defined on } \omega \text{ such that } f(0) = 0, \\ f(k+1) = \mathscr{P}(f(k)) \text{ for all } k \in \omega, \text{ and } n \in \omega \text{ and } x \in f(n)].$$

This fact, coupled with the relative consistency proof of *AC* just given, yields the following result.

THEOREM: *If an arithmetical formula E (one whose quantifiers are relativized to V_ω, that is) is derivable in the theory ZF+AC, it is derivable already in ZF.*

PROOF: Since $(x \in V_\omega)^{HOD} \leftrightarrow x \in V_\omega$ we can see without difficulty that $E^{HOD} \leftrightarrow E$, if E has been relativized to V_ω.

So if E is a theorem of $ZF+AC$, there is a proof $A_0, \ldots, A_{n-2}, E$ of it in that theory. We have shown, however, for each axiom A of $ZF+AC$ that A^{HOD} is a theorem of ZF. Thus $A_0^{HOD}, \ldots, A_{n-2}^{HOD}, E^{HOD}$ is a proof of E^{HOD}, and so of E, from ZF alone. ∎

The Choice Principle: We say that the *choice principle* holds for a universe $\mathscr{U}$ if there is a formula $A(x, y)$ of two free variables and no parameters which defines a well-ordering of the entire universe.

In such a case the axiom of choice is easily seen to hold. We should note however that, as it stands, the choice principle is not only not an axiom, it is not even a scheme of axioms; rather, it is something like an 'infinite disjunction' of sentences, of all those sentences, that is, which say, of a binary relation without parameters, that it well-orders the universe $\mathscr{U}$. Nevertheless, we can establish the following result.

THEOREM: *If AF holds in $\mathscr{U}$, then the choice principle holds in $\mathscr{U}$ if and only if the axiom $\forall x OD(x)$ holds in $\mathscr{U}$.*

PROOF: If the choice principle holds, some parameter-free formula $A(x, y)$ defines a well-ordering of $\mathscr{U}$. So there will be generated by this well-ordering an isomorphism $x = J(\alpha)$ from On on to $\mathscr{U}$, and this isomorphism will be definable by a formula free of parameters. Thus $x = J(\alpha)$ will provide an ordinal definition for every element of $\mathscr{U}$, and therefore $\forall x OD(x)$ will hold.

Conversely, given that $\forall x OD(x)$ holds, we can proceed as above to write down the formula $D(x) \leqslant D(y)$; this is parameter-free, and it defines a well-ordering on $\mathscr{U}$. ▮

In particular, granted *AF*, the parameter-free formula $D(x) \leqslant D(y)$ has the following property. If there is any parameter-free formula $A(x, y)$ defining a well-ordering of the universe, then $D(x) \leqslant D(y)$ defines a well-ordering of the universe.

In Chapter VIII we will be able to show the relative consistency of the axiom $\forall x OD(x)$.

CHAPTER VII

FRAENKEL/MOSTOWSKI MODELS

Relative Consistency of the Negation of the Axiom of Choice (Without the Axiom of Foundation)

Let $R(x, y)$ be a relation of two arguments that establishes a bijection from the universe $\mathcal{U}$ on to itself; that is, a relation which satisfies the following conditions:

$$\forall x \forall y \forall y' [R(x, y) \wedge R(x, y') \rightarrow y = y'];$$
$$\forall x \forall x' \forall y [R(x, y) \wedge R(x', y) \rightarrow x = x'];$$
$$\forall x \exists y R(x, y) \wedge \forall y \exists x R(x, y).$$

We shall write this functional relation as $y = F(x)$; and the binary relation $x \in F(y)$ will be written $x \in' y$. Likewise, the formula obtained from the formula $E(x_0, \ldots, x_{k-1})$ by changing $\in$ throughout to $\in'$ will be written $E'(x_0, \ldots, x_{k-1})$.

We shall show that every ZF axiom holds in the class of all sets, enriched with the relation $\in'$ (a universe we shall call $\mathcal{U}'$). This amounts to proving that if A is an axiom of ZF, then A' holds in $\mathcal{U}$.

Axiom of Extensionality: We must show that

$$\forall x \forall y [x = y \leftrightarrow \forall z (z \in' x \leftrightarrow z \in' y)].$$

So let a, b be two sets in $\mathcal{U}$ such that $\forall z (z \in' a \leftrightarrow z \in' b)$; by the definition of $\in'$ we then have

$$\forall z (z \in F(a) \leftrightarrow z \in F(b)).$$

Thus $F(a) = F(b)$. Since F is bijective, this means that $a = b$.

Union Axiom: If a is a set then the class $\exists y [y \in' a \wedge x \in' y]$ is one too. For this formula is just an abbreviation of $\exists y [y \in F(a) \wedge x \in F(y)]$, and so is equivalent to $x \in \bigcup_{y \in F(a)} F(y)$. Putting this last set equal to c, we obtain

$$\forall x [x \in c \leftrightarrow \exists y (y \in' a \wedge x \in' y)],$$

whence, letting b stand for $F^{-1}(c)$,

$$\forall x [x \in' b \leftrightarrow \exists y (y \in' a \wedge x \in' y)],$$

which shows that the union axiom holds.

Power-set Axiom: If a is any set, the formula $\forall y(y \in' x \rightarrow y \in' a)$ is just

$$\forall y\,[y \in F(x) \rightarrow y \in F(a)];$$

that is, it is equivalent to $F(x) \subset F(a)$ or, putting c for $\mathscr{P}(F(a))$, to $F(x) \in c$. Defining b by $F(b) = \{x \mid F(x) \in c\}$, and remembering that F is bijective, we get $F(x) \in c$ equivalent to $x \in F(b)$. Thus

$$x \in' b \leftrightarrow \forall y\,[y \in' x \rightarrow y \in' a],$$

proving that the power-set axiom also holds.

Scheme of Replacement: Let a be a set and $R(x, y)$ a formula such that $R'(x, y)$ defines a functional relation. Let c be the set of images of elements of $F(a)$ under this functional relation. Then

$$\forall y\,[y \in c \leftrightarrow \exists x(x \in F(a) \wedge R'(x, y))];$$

so by writing again b for $F^{-1}(c)$ we have

$$\forall y\,[y \in' b \leftrightarrow \exists x(x \in' a \wedge R'(x, y))].$$

Thus the replacement scheme holds.

Axiom of Infinity: We define by recursion a map f with domain $\omega : f(0) = F^{-1}(0)$; $f(n+1)$ is defined by $F(f(n+1)) = F(f(n)) \cup \{f(n)\}$. Writing η for the range of f, we then set $\theta = F^{-1}(\eta)$.

Since $\forall x(x \notin' F^{-1}(0))$, the set $F^{-1}(0)$ must be the empty set $\emptyset'$ of $\mathscr{U}'$. Since $\emptyset' \in \eta$, we get $\emptyset' \in' \theta$.

Moreover, if $x \in' \theta$, then $x \in \eta$, so $x = f(n)$ for some $n \in \omega$. Now $\forall z[z \in' f(n+1) \leftrightarrow z \in' f(n) \vee z = f(n)]$, from the definition of $f(n+1)$; consequently,

$$f(n+1) = x \cup' \{x\}',$$

and so $\forall x[x \in' \theta \rightarrow x \cup' \{x\}' \in' \theta]$; thus θ is a set satisfying the axiom of infinity in $\mathscr{U}'$.

If the axiom of choice holds in $\mathscr{U}$, it holds also in $\mathscr{U}'$.

For suppose that a is a set for which, in $\mathscr{U}'$, the formula 'the elements of a are non-empty but pairwise disjoint' holds. Then

$$\forall x\,[x \in' a \rightarrow \exists y(y \in' x)]$$

and

$$\forall x \forall y\,[(x \in' a \wedge y \in' a \wedge x \neq y) \rightarrow \forall z(z \notin' x \vee z \notin' y)].$$

Put a_1 for $\{F(x) \mid x \in F(a)\}$; then the two sentences above say that the elements of a_1 are non-empty and disjoint in pairs. So if the axiom of choice holds in $\mathscr{U}$, we can find a set b_1 which intersects each member of a_1 at a single point; that is,

$$\forall x \exists y [x \in a_1 \rightarrow \forall z (y = z \leftrightarrow z \in x \wedge z \in b_1)],$$

or, alternatively,

$$\forall x \exists y [F(x) \in a_1 \rightarrow \forall z (y = z \leftrightarrow z \in F(x) \wedge z \in b_1)].$$

Now the definition of a_1 implies that $F(x) \in a_1$ iff $x \in F(a)$; if and only if, that is, $x \in' a$. Thus we get

$$\forall x \exists y [x \in' a \rightarrow \forall z (y = z \leftrightarrow z \in' x \wedge z \in b_1)].$$

Write now b for $F^{-1}(b_1)$; we obtain

$$\forall x \exists y [x \in' a \rightarrow \forall z (y = z \leftrightarrow z \in' x \wedge z \in' b)],$$

showing that the axiom of choice holds in $\mathscr{U}'$.

A set a is called an atom of the universe $\mathscr{U}$ if $a = \{a\}$; or (equivalently) if $\forall x (x \in a \leftrightarrow x = a)$ holds in $\mathscr{U}$. No atoms can exist, it is clear, where the axiom of foundation holds.

If ZF is consistent so is ZF+'there is an atom'.

To prove this we define a simple bijection F from $\mathscr{U}$ on to itself by the formula $(x=0 \wedge y=1) \vee (x=1 \wedge y=0) \vee (x \neq 0 \wedge x \neq 1 \wedge x=y)$. In the universe $\mathscr{U}'$ obtained by applying this bijection to $\mathscr{U}$, the empty set of $\mathscr{U}$, namely $\emptyset$, is an atom. For since $F(\emptyset) = \{\emptyset\}$, we have at once $\forall x (x \in' \emptyset \leftrightarrow x = \emptyset)$.

It is possible, by a suitable choice for the bijection F, to show that several other sentences, stronger than the negation of *AF*, are consistent with *ZF*; for example, the sentences asserting the existence of $\in$-cycles $a_0 \in a_1 \in \ldots \in a_{n-1} \in a_0$.

For the rest of this chapter we shall rely on the following example only.

If ZF is consistent, so is ZF+'there exists a set of atoms that is equipollent to ω'.

Define F by setting $F(n) = \{n\}$, $F(\{n\}) = n$ for every natural number $n \geqslant 1$, and $F(x) = x$ everywhere else. This is a bijection all right since n and $\{p\}$ must be different if n and p are both positive natural numbers.

Since for every $n \geqslant 1$ we have $\forall x(x \in' n \leftrightarrow x = n)$ we see that n is an atom of $\mathscr{U}'$. We must now show that, within $\mathscr{U}'$, this set of atoms is equipollent to ω.

Given two objects a, b we write (we have already done this above, without explanation) $\{a, b\}'$ for their pair in $\mathscr{U}'$; it is clear that $\{a, b\}' = F^{-1}(\{a, b\})$. $\langle a, b\rangle'$ is similarly the ordered pair in $\mathscr{U}'$ of a and b, so that $\langle a, b\rangle' = F^{-1}(\{F^{-1}(\{a\}), F^{-1}(\{a, b\})\})$.

We can recursively define a function f with domain ω by setting

$$\forall x [x \in' f(n) \leftrightarrow \exists i < n(x = f(i))];$$

thus $f(n+1) = F^{-1}(\{f(0), f(1), \ldots, f(n)\})$ and $f(0) = \emptyset'$. From this definition it is clear that when n runs through ω, $f(n)$ runs through the set of natural numbers in $\mathscr{U}'$.

By putting $\langle x, y\rangle' \in' g \leftrightarrow \langle x, y\rangle \in f$, we can then define a map g in $\mathscr{U}'$. We obtain $\langle x, y\rangle' \in F(g) \leftrightarrow \langle x, y\rangle \in f$, and thus if h is the set of images of elements of f under the map $\langle x, y\rangle \to \langle x, y\rangle'$, we have $g = F^{-1}(h)$.

After all this, it is clear that in $\mathscr{U}'$ g is a bijection from the natural numbers of $\mathscr{U}$ to the natural numbers of $\mathscr{U}'$. Since all the positive natural numbers of $\mathscr{U}$ are atoms in $\mathscr{U}'$, we have shown that the latter universe contains a set of atoms equipollent to ω.

RELATIVE CONSISTENCY OF THE NEGATION OF THE AXIOM OF CHOICE

In this section we shall provide a model of ZF in which the axiom of choice fails. From what we have just proved we can suppose given a universe $\mathscr{U}_0$ in which there is a set of atoms, X, equipollent to ω.

We begin by defining a functional relation $y = W_\alpha$ on the ordinals by $W_0 = X$; $W_\alpha = \bigcup_{\beta<\alpha} \mathscr{P}(W_\beta)$ (for all $\alpha \neq 0$).

Note that since X is a set of atoms, W_0 is transitive. Thus $W_0 \subset \mathscr{P}(W_0) = W_1$. For $1 \leqslant \beta \leqslant \alpha$, furthermore, it is clear that $W_\beta \subset W_\alpha$. It follows that whatever ordinals α, β are, if $\beta \leqslant \alpha$ then $W_\beta \subset W_\alpha$.

In the same way as we proved similar results for V_α in Chapter III, we can show that $W_{\alpha+1} = \mathscr{P}(W_\alpha)$, and $W_\alpha = \bigcup_{\beta<\alpha} W_\beta$ when α is a limit ordinal. The union (in the intuitive sense) of all the W_α we denote by W; $W(x)$, then, is the formula $\exists\alpha[On(\alpha) \wedge x \in W_\alpha]$.

If a is in W we again call the least ordinal α for which $a \in W_\alpha$ the *rank* of a.

LEMMA: *An object a is in W if and only if all its elements are. The rank of any element of a is strictly below that of a itself, provided only that a does not have rank* 0. *The axioms of ZF hold in the class W.*
PROOF: All parts of the lemma are proved in the same way as similar results were proved for V in Chapter III. ▮

The set X is in W since $X \in W_1$. Moreover, every atom in W is an element of X. For suppose, on the contrary, that there were an atom a of W whose rank were non-zero. Since no limit ordinal is the rank of anything, we can suppose a's rank to be $\alpha+1$. Thus $a \in W_{\alpha+1}$, so $a \subset W_\alpha$; but $a \in a$, so $a \in W_\alpha$, contradicting the definition of rank. Thus a has rank 0, so is in X.

Again, the bijection f from X on to ω is in W; for f consists only of pairs $\langle a, n\rangle$ with $a \in X$ and $n \in \omega$; $\langle a, n\rangle$ is therefore in W_{n+3}.

Although W does not satisfy the axiom of foundation, it does satisfy the sentence

$$\forall x[x \neq \emptyset \rightarrow \exists y[y \in x \wedge (y \cap x = \emptyset \vee y = \{y\})]].$$

For if a non-empty set a is in W, let b be an element of it of least rank. If the rank in question is positive, every element of b is of lower rank, so $b \cap a = \emptyset$; but if it is zero, b is an atom.

Consequently the class W proves the consistency of the axiom system T given by: ZF+*'the class of all atoms (defined by the formula* $x=\{x\}$*) is a set equipollent to* ω*'*$+\forall x[x\neq\emptyset \rightarrow \exists y[y \in x \wedge (y \cap x=\emptyset \vee y= =\{y\})]]$.

We now consider a universe $\mathscr{U}_1$ satisfying all these axioms; and, writing X for the set of atoms of $\mathscr{U}_1$ we carry out the above construction within $\mathscr{U}_1$. Thus $W_0=X$ and if $\alpha\geqslant 1$, $W_\alpha = \bigcup_{\beta<\alpha} \mathscr{P}(W_\beta)$; W is the class defined by $\exists\alpha[On(\alpha) \wedge x \in W_\alpha]$.

Every object of $\mathscr{U}_1$ is in W; equivalently, $\mathscr{U}_1$ *satisfies the axiom* $\forall x\exists\alpha[On(\alpha) \wedge x \in W_\alpha]$. For suppose the existence of an a not in W; let b be its transitive closure, and c the set of those elements of b not in W. Since a is not in W, some element x of a is not in W; and since this x will be in c, c cannot be empty. So take any $y \in c$; it is not in W, so has an element z not in W. Since $y \in b$ and b is transitive, z is also in b. Thus z qualifies for c, which means that $y \cap c \neq 0$, for every $y \in c$. Moreover, no such y can be an atom either, for in this universe all atoms are in W. The set c thus contradicts the axiom

$$\forall x [x \neq \emptyset \rightarrow \exists y [y \in x \wedge (y \cap x = \emptyset \vee y = \{y\})]].$$

If σ is a bijection from X on to itself (a *permutation* it is also called), we can define by recursion a functional relation σ_α such that (1) if α is an ordinal σ_α is an $\in$-automorphism of W_α and (2) $\sigma_0 = \sigma$, and if $\beta' \leqslant \beta$ then σ_β is an extension of $\sigma_{\beta'}$. For suppose the family $(\sigma_\beta)_{\beta<\alpha}$ has been defined to meet these requirements. Then for limit α we have $W_\alpha = \bigcup_{\beta<\alpha} W_\beta$, so we can define σ_α simply as $\bigcup_{\beta<\alpha} \sigma_\beta$; and the proof that σ_α has properties (1) and (2) is entirely straightforward.

For $\alpha = \beta + 1$ we have $W_\alpha = \mathscr{P}(W_\beta)$, so can define σ_α by setting $\sigma_\alpha(u) = \{\sigma_\beta(x) \mid x \in u\}$ for every $u \in W_\alpha$. Since σ_β is a permutation of W_β, it is clear that σ_α is a permutation of $\mathscr{P}(W_\beta)$. In addition, σ_α is an extension of σ_β; for if $u \in W_\beta$ each $x \in u$ belongs to W_β, and since σ_β is an $\in$-automorphism of W_β, we have $x \in u \leftrightarrow \sigma_\beta(x) \in \sigma_\beta(u)$, so that $\sigma_\beta(u) = \{\sigma_\beta(x) \mid x \in u\} = \sigma_\alpha(u)$.

Finally, σ_α is itself an automorphism of W_α, as can be shown by proving that, for arbitrary $x, u \in W_\alpha$,

$$x \in u \leftrightarrow \sigma_\alpha(x) \in \sigma_\alpha(u).$$

But if $x \in u$, we get $x \in W_\beta$, so $\sigma_\beta(x) \in \sigma_\alpha(u)$ – by the definition of σ_α –, so

$$\sigma_\alpha(x) \in \sigma_\alpha(u).$$

Conversely, if $\sigma_\alpha(x) \in \sigma_\alpha(u)$, then $\sigma_\alpha(x) \in W_\beta$. As σ_α is a permutation of W_α which simply extends the permutation σ_β of W_β, we have $x \in W_\beta$, so $\sigma_\beta(x) \in \sigma_\alpha(u)$, and therefore $x \in u$, by the definition of σ_α.

Starting from $\sigma_0 = \sigma$ then, the functional relation $\alpha \rightarrow \sigma_\alpha$ as recursively defined above has all the required properties.

From it we can define another functional relation, $y = \mathscr{S}_\sigma(x)$, as the common extension of all the σ_α; this functional relation, defined by

$$\exists \alpha [On(\alpha) \wedge x \in W_\alpha \wedge y = \sigma_\alpha(x)],$$

has domain $\mathscr{U}_1$ and, indeed, is an automorphism of $\mathscr{U}_1$, so that the following are all true in $\mathscr{U}_1$:

$$\forall x \forall x' [\mathscr{S}_\sigma(x) = \mathscr{S}_\sigma(x') \to x = x'];$$
$$\forall y \exists x [y = \mathscr{S}_\sigma(x)];$$
$$\forall x \forall x' [x \in x' \leftrightarrow \mathscr{S}_\sigma(x) \in \mathscr{S}_\sigma(x')].$$

(The proof of these sentences is immediate from the definition of $\mathscr{S}_\sigma$.)

Thus for each permutation σ of the set X of atoms we have defined a functional relation $y = \mathscr{S}_\sigma(x)$ which is an automorphism of the entire universe $\mathscr{U}_1$. For notational simplicity we shall from here on write $y = \sigma(x)$ for $y = \mathscr{S}_\sigma(x)$; since for any atom x we have $\sigma(x) = \mathscr{S}_\sigma(x)$, this convention will not precipitate any confusion.

LEMMA: *Let $E(a_0, \ldots, a_{n-1})$ be a sentence with parameters $a_0, \ldots, a_{n-1}$; and let σ be a permutation of X. Then $E(a_0, \ldots, a_{n-1}) \leftrightarrow E(\sigma a_0, \ldots, \sigma a_{n-1})$.*
PROOF: The proof is by informal induction on the length of the formula $E(a_0, \ldots, a_{n-1})$; if this formula is atomic, that is, $a_0 \in a_1$ or $a_0 = a_1$, then the theorem holds because σ is an $\in$-automorphism of the universe $\mathscr{U}_1$.

If $E(a_0, \ldots, a_{n-1})$ is $\neg F(a_0, \ldots, a_{n-1})$ we have $F(a_0, \ldots, a_{n-1}) \leftrightarrow F(\sigma a_0, \ldots, \sigma a_{n-1})$ by the induction hypothesis; and therefore the required equivalence holds also for E. When $E(a_0, \ldots, a_{n-1})$ is $F(a_0, \ldots, a_{n-1}) \vee \vee\, G(a_0, \ldots, a_{n-1})$ the proof is the same.

If $E(a_0, \ldots, a_{n-1})$ is $\exists x F(x, a_0, \ldots, a_{n-1})$, suppose it to be true in $\mathscr{U}_1$; then for some object a we have $F(a, a_0, \ldots, a_{n-1})$, and so $F(\sigma a, \sigma a_0, \ldots, \sigma a_{n-1})$ by the induction hypothesis; thus $\exists x F(x, \sigma a_0, \ldots, \sigma a_{n-1})$. The converse is handled likewise. ▮

Note that this lemma is another example of a theorem scheme, this time in the theory T defined on p. 74 above. What we have shown is the sentence 'For every permutation σ of the set of atoms, and for every $x_0, \ldots, x_{n-1}$, $E(x_0, \ldots, x_{n-1}) \leftrightarrow E(\sigma x_0, \ldots, \sigma x_{n-1})$' for an arbitrary parameter-free formula $E(x_0, \ldots, x_{n-1})$.

LEMMA: *Let $A(x, a_0, \ldots, a_{k-1})$ be a formula of one free variable holding for just one set, the set a; let σ be a permutation of X. Then σa is the only set for which the formula $A(x, \sigma a_0, \ldots, \sigma a_{k-1})$ holds.*
PROOF: By assumption we have $\forall x [x = a \leftrightarrow A(x, a_0, \ldots, a_{k-1})]$; so, by the previous lemma,

$$\forall x [x = \sigma a \leftrightarrow A(x, \sigma a_0, \ldots, \sigma a_{k-1})]. ▮$$

LEMMA: *If α is an ordinal, and σ is a permutation of X, then $\sigma\alpha=\alpha$.*
PROOF: Otherwise, suppose α is the first ordinal for which $\sigma\alpha\neq\alpha$; then $\sigma\beta=\beta$ for all $\beta\in\alpha$, so $\sigma\alpha=\{\sigma\beta \mid \beta\in\alpha\}=\alpha$, which contradicts the supposition. ▌

Within the universe $\mathscr{U}_1$ we can apply the generalized reflection principle (p. 51) to the functional relation $y=W_\alpha$. Since the universe is exhausted in this case by the (intuitive) union of all the W_α's, we can conclude that for every formula E that is without parameters

$$\forall\alpha\exists\beta>\alpha\forall x_0\ldots\forall x_{k-1}\left[x_0\in W_\beta\wedge\ldots\wedge x_{k-1}\in W_\beta \rightarrow\left(E(x_0,\ldots,x_{k-1})\leftrightarrow E^{W_\beta}(x_0,\ldots,x_{k-1})\right)\right].$$

We can proceed to define the class OD_1 by the formula $OD_1(x)$, 'There are an ordinal α and a monadic expression $\Phi(y,\alpha_0,\ldots,\alpha_{r-1},a_0,\ldots,a_{s-1})$ all of whose parameters are atoms or ordinals $<\alpha$, and $\mathrm{val}(\Phi, W_\alpha)=\{x\}$'.

This class will be referred to as the class of *sets definable by ordinals and atoms*. As in Chapter VI we can prove the following result.

LEMMA: *Let $A(x,\alpha_0,\ldots,\alpha_{r-1},u)$ be a singulary formula whose parameters $\alpha_0,\ldots,\alpha_{r-1},u$ are all ordinals, except for u, which is a finite sequence of atoms (a map, that is to say, from some natural number s into X). Then if a is the only set for which the formula A holds, a is definable by ordinals and atoms.*
PROOF: The generalized reflection principle provides us with an ordinal $\alpha>\alpha_0,\ldots,\alpha_{r-1}$ and big enough for u and a both to belong to W_α, and such that W_α mirrors the formula A.

Then the value of the expression $\ulcorner A(x,\alpha_0,\ldots,\alpha_{r-1},u)\urcorner$ in W_α is $\{a\}$. Write $\Psi(x,\alpha_0,\ldots,\alpha_{r-1},u)$ for this expression. Since u is a finite sequence $a_0,\ldots,a_{s-1}$ of atoms we have

$$u=\{\langle 0,a_0\rangle,\ldots,\langle s-1,a_{s-1}\rangle\}.$$

Thus the expression

$$\exists z(\forall t[t\,\varepsilon\, z\Leftrightarrow t=\langle 0,a_0\rangle\vee\ldots\vee t=\langle s-1,a_{s-1}\rangle]\wedge\Psi(x,\alpha_0,\ldots,\alpha_{r-1},z))$$

gets the value $\{a\}$ in W_α and has only ordinals and atoms for parameters. ▌

The converse can also be obtained.

CONVERSE: *If a is a set definable by ordinals and atoms, then there is a formula $A(x, \delta_0, u)$ of one free variable that holds for a and a alone; the parameters of A, moreover, are just two – the ordinal δ_0 and the finite sequence of atoms u.*

PROOF: By the specification of a we have an ordinal γ_0 and an expression $\Phi(x, \alpha_0, \ldots, \alpha_{r-1}, a_0, \ldots, a_{s-1})$ whose parameters are the atoms $a_0, \ldots, a_{s-1}$ and the ordinals $\alpha_0, \ldots, \alpha_{r-1} < \gamma_0$; and the value of Φ in W_{γ_0} is $\{a\}$.

Such an expression specifies uniquely a natural number (the one associated with the parameter-free expression $\Phi(x, x_0, \ldots, x_{r-1}, y_0, \ldots, y_{s-1})$), an ordinal (the one associated with the finite sequence of ordinals $\alpha_0, \ldots, \alpha_{r-1}$), and a finite sequence $u = a_0, \ldots, a_{s-1}$ of atoms; and therefore it singles out an ordinal β_0 and a finite sequence of atoms, u. A formula with three parameters $E(x, \gamma_0, \beta_0, u)$, namely 'the expression (whose parameters are ordinals and atoms) associated with the ordinal β_0 and the finite atom-sequence u has the value $\{x\}$ in the set W_{γ_0}' is therefore at hand. The number of parameters can then be reduced to two by means of the bijection from On^2 on to On; and this yields a formula $A(x, \delta_0, u)$ holding, as required, for a and nothing else. Such a formula $A(x, \delta_0, u)$ will be said to be a *definition of a by ordinals and atoms.* ▮

Note that the class OD_1 has been defined by a parameter-free formula. Also parameter-free is the formula $HOD_1(x)$ which defines the class of *sets hereditarily definable by ordinals and atoms*, HOD_1. It is simply 'Every element of $\mathscr{C}(\{x\})$ is in OD_1'.

LEMMA: *A set a is in HOD_1 if and only if a is in OD_1 and every element of a is in HOD_1.*

PROOF: The proof is the same as that of the corresponding lemma in Chapter VI. ▮

That *the class HOD_1 satisfies the axioms of ZF* can also be proved in the way the same claim for HOD was proved in the last chapter.

We can prove that the set X of atoms is in HOD_1. For every atom a is definable by the formula $x = a$, so is in OD_1. As $\mathscr{C}(\{a\}) = a$, each atom is in HOD_1. So we need only show that X is in OD_1; but since it is the set of atoms, it is definable by $\forall z(z \in x \leftrightarrow z = \{z\})$, so certainly qualifies for OD_1.

But *the axiom of choice is not satisfied in HOD_1*; worse, *the set of atoms cannot even be linearly ordered.*

For suppose in HOD_1 there were a set v which imposed a strict linear order on X. Let $A(x, \delta_0, u)$ be some definition of v by ordinals and atoms. Then since u is only a finite sequence of atoms $a_0, \ldots, a_{s-1}$, and X is equipollent with ω (in $\mathscr{U}_1$), there must be two atoms b, c different from all the $a_0, \ldots, a_{s-1}$. v linearly orders X, however, so without loss of generality we can suppose $\langle b, c\rangle \in v$. Let σ be that permutation which switches b and c round, but otherwise leaves X unaltered; then nothing happens to any of the atoms $a_0, \ldots, a_{s-1}$ under σ, so $\sigma u = u$. Moreover, $\sigma\delta_0 = \delta_0$, since δ_0 is an ordinal. But, by the lemma on p. 76 σv is defined by the formula $A(x, \sigma\delta_0, \sigma u)$, so σv actually equals v. Now $\langle b, c\rangle \in v$ by assumption, so $\langle \sigma b, \sigma c\rangle \in \sigma v$, so $\langle c, b\rangle \in v$. Thus v is not a strict order, which contradicts our hypothesis.

Let us note two other properties of X which contradict AC and yet are true of X in HOD_1.

Every subset of X either is finite or has a finite complement with respect to X.

Let Y be some object of HOD_1 which is a subset of X, but not finite (that is to say, it is not equipollent with any natural number); and suppose $A(x, \gamma_0, u)$ to be a definition of Y in terms of ordinals and atoms. As Y is not finite we can certainly find a $b \in Y$ which is not among the terms of the finite atom-sequence u. Let c be any other atom ignored by u, and let σ be that permutation of X which switches b and c round whilst keeping the other atoms fixed. We have at once that $\sigma u = u$ and $\sigma\gamma_0 = \gamma_0$ (since γ_0 is an ordinal). And as Y is defined by the formula $A(x, \gamma_0, u)$, we must also have $\sigma Y = Y$. As $b \in Y$ we get $\sigma b \in \sigma Y$, so $c \in Y$. Thus every atom like c, that is every atom not accounted for in u, is an element of Y. Thus $X \setminus Y$ is finite.

X is not finite, but it has no infinite denumerable subset (subset equipollent with ω).

As X is equipollent with ω within $\mathscr{U}_1$, it cannot in HOD_1 be equipollent to a natural number. If Y is a subset of X that is equipollent with ω, it cannot be finite; and so $X \setminus Y$ is finite according to the previous result. But then X itself is equipollent with ω, which means that it can be well ordered. And this, we know, is false.

Note that the model of $ZF + \neg AC$ just constructed does not satisfy

AF, so it does not solve the problem of whether $ZF + AF + \neg AC$ is consistent. Indeed, the set here which cannot be well ordered is a pathological one, the set of atoms. The results of P. Cohen described in [2] yield far more interesting relative consistency results; for example that of $ZF + AF +$ '$\mathscr{P}(\omega)$ cannot be well ordered'.

CHAPTER VIII

CONSTRUCTIBLE SETS

Relative Consistency of the Generalized Continuum Hypothesis

Given a set X and a subset Y of it we say that Y is a *subset of X definable by parameters*, if there is a monadic expression $\Phi(x, a_0, \ldots, a_{k-1})$ which has parameters $a_0, \ldots, a_{k-1}$ in X and the value Y in the set X.

We define a functional relation $y = \prod(x)$ on the class of all sets by the formula 'y is the set of subsets of x definable by parameters'.

If the axiom of choice holds, then if a is an infinite set, $\overline{\overline{\prod(a)}} = \bar{\bar{a}}$.

For, first of all, every member b of a provides a subset of a, $\{b\}$, definable by parameters by way of the expression $x = b$; so $\overline{\overline{\prod(a)}} \geqslant \bar{\bar{a}}$. And the set of expressions with parameters in a has cardinality $\aleph_0 \cdot \overline{\overline{\sigma(a)}} = \bar{\bar{a}}$ (since any such expression can be specified by a parameter-free expression and a finite sequence of members of a); so $\overline{\overline{\prod(a)}} \leqslant \bar{\bar{a}}$.

This proves that for infinite a, $\prod(a)$ is a proper subset of $\mathscr{P}(a)$.

Observe however that the functional relation $y = \prod(x)$ is not increasing with x; that, in other words, we can have $a \subset b$ without having $\prod(a) \subset \prod(b)$. For if a is a subset of b, but not one definable by parameters, then $a \in \prod(a)$ (defined as it is by the expression $x = x$), but $a \notin \prod(b)$. However, if a is also a member of b, this possibility no longer obtains.

THEOREM: *If $a \subset b$ and $a \in b$ then $\prod(a) \subset \prod(b)$.*

PROOF: If $c \in \prod(a)$ then $c = \mathrm{val}(\Phi, a)$ for some monadic expression $\Phi(x)$ whose parameters $a_0, \ldots, a_{k-1}$ are in a. Thus $z \in c$ iff $z \in \mathrm{val}(\Phi, a)$, that is iff $\{\langle x, z\rangle\} \in \mathrm{Val}(\Phi, a)$. But by the theorem on pp. 61f., $\mathrm{Val}(\Phi, a) = \mathrm{Val}(\Phi^a, b) \cap {}^{\{x\}}a$; thus $z \in c$ iff $\{\langle x, z\rangle\} \in \mathrm{Val}(\Phi^a, b) \wedge \{\langle x, z\rangle\} \in {}^{\{x\}}a$. Consequently,

$$z \in c \leftrightarrow z \in \mathrm{val}(\Phi^a, b) \wedge z \in a,$$

so $c = \mathrm{val}(\Phi^a, b) \cap a$.

It follows at once that $c = \mathrm{val}[\Phi^a(x) \wedge x \varepsilon a, b]$, since in general $\mathrm{val}(\Phi, b)$ is the set of elements of b that satisfy Φ. But all the parameters of the expression $\Phi^a(x) \wedge x \varepsilon a$, namely $a, a_0, \ldots, a_{k-1}$ are elements of b. Thus $c \in \prod(b)$. ▮

By setting first $L_0 = \emptyset$ we can define recursively a functional relation on the whole of *On* by setting thereafter

$$L_\alpha = \bigcup_{\beta<\alpha} \prod(L_\beta).$$

The class L defined by the formula $L(x) : \exists\alpha[On(\alpha) \wedge x \in L_\alpha]$, the union (in the intuitive sense) of all the L_α, is called the class of *constructible sets*. Appropriately enough, a set a is called *constructible* if it is in L; if, therefore, it is in L_α for some ordinal α.

The axiom of constructibility is the sentence 'Every set is constructible', or, equivalently, $\forall x \exists\alpha[On(\alpha) \wedge x \in L_\alpha]$. The main business of this chapter will consist in showing that if *ZF* is consistent, it remains so on the addition of the axiom of constructibility. We do this by proving that *the class L with the usual membership relation satisfies all the ZF axioms and the axiom of constructibility too.*

We first prove some results for the L_α's very similar to ones proved already for the V_α's. If $\alpha' \leqslant \alpha$ then $L_{\alpha'} \subset L_\alpha$, since $L_{\alpha'} = \bigcup_{\beta<\alpha'} \prod(L_\beta) \subset \bigcup_{\beta<\alpha} \prod(L_\beta) = L_\alpha$. If $\beta < \alpha$, then $L_\beta \in L_\alpha$; for $L_\beta \in \prod(L_\beta)$ and $\prod(L_\beta) \subset L_\alpha$. By our most recent theorem then, since $L_\beta \subset L_\alpha$ and $L_\beta \in L_\alpha$, we have $\prod(L_\beta) \subset \prod(L_\alpha)$.

As $L_{\alpha+1} = \bigcup_{\beta\leqslant\alpha} \prod(L_\beta)$ we have $L_{\alpha+1} = \prod(L_\alpha)$ *for every ordinal* α. On the other hand, if α is a limit ordinal,

$$L_\alpha = \bigcup_{\beta<\alpha} \prod(L_\beta) = \bigcup_{\beta<\alpha} L_{\beta+1},$$

so that $L_\alpha = \bigcup_{\beta<\alpha} L_\beta$ *for every limit ordinal* α.

Corresponding to the definition of *rank*, we now provide a definition of *order*. The *order* of a constructible set x is the first ordinal α for which $x \in L_\alpha$. We write it $\mathrm{od}(x)$, and note that $\mathrm{od}(x)$ is never a limit ordinal; for if α is a limit and $x \in L_\alpha$ then x is already in L_β for some smaller β.

LEMMA: *If a is constructible, so are all its elements; their orders, moreover, are strictly less than the order of a.*
PROOF: Let $\alpha=\beta+1$ be the order of a. Then $a \in L_{\beta+1}$, so $a \in \prod(L_\beta)$. Thus $a \subset L_\beta$, and every element of a must belong to L_β; these elements, therefore, have lower order than a. ▮

It follows that for any ordinal α the set L_α is transitive; for if $a \in L_\alpha$ and $b \in a$, the order of a is $\leqslant \alpha$, so also the order of b.

THEOREM: *Every ordinal α is constructible, and the order of α is $\alpha+1$.*
PROOF: If there is an ordinal α for which $\alpha \in L_\alpha$, let α_0 be the least. Then $\alpha_0 \in \bigcup_{\beta \in \alpha_0} \prod(L_\beta)$ and so $\alpha_0 \in \prod(L_\beta)$ for some $\beta \in \alpha_0$; thus $\alpha_0 \subset L_\beta$, giving $\beta \in L_\beta$, in contradiction with the definition of α_0.

That much done, it remains to prove that $\alpha \in L_{\alpha+1}$ for every ordinal α. Again, let α_0 be the least ordinal, if there is one, such that $\alpha_0 \notin L_{\alpha_0+1}$.

For every $\beta \in \alpha_0$ then, we have $\beta \in L_{\beta+1}$, so $\beta \in \prod(L_\beta)$. It follows that $\alpha_0 \subset \bigcup_{\beta<\alpha_0} \prod(L_\beta)$; and therefore that $\alpha_0 \subset L_{\alpha_0}$.

Consider now the parameter-free expression $\Phi(x)$ of one free variable

$$\forall u \forall v [u \varepsilon x \wedge v \varepsilon x \Rightarrow u \varepsilon v \vee u = v \vee v \varepsilon u] \wedge \forall u \forall v [u \varepsilon x \wedge v \varepsilon u \Rightarrow v \varepsilon x].$$

Let the value of this expression in the set L_{α_0} be Y, which is thus the set of elements x of L_{α_0} for which the following formula, which is relativized to L_{α_0}, holds.

$$\begin{aligned} &\forall u \forall v [u \in L_{\alpha_0} \wedge v \in L_{\alpha_0} \to [(u \in x \wedge v \in x) \\ &\quad \to (u \in v \vee u = v \vee v \in u)]] \\ &\quad \wedge \forall u \forall v [u \in L_{\alpha_0} \wedge v \in L_{\alpha_0} \to ((u \in x \wedge v \in u) \to v \in x)]. \end{aligned}$$

Seeing that L_{α_0} is transitive, an element x of it will satisfy this formula if and only if

$$\forall u \forall v [u \in x \wedge v \in x \to u \in v \vee u = v \vee v \in u] \wedge \forall u \forall v [u \in x \wedge v \in u \to v \in x];$$

just, that is, when it is an ordinal.

The value of $\Phi(x)$ in L_{α_0} is therefore the set of ordinals belonging to L_{α_0}. We have proved that α_0 anyway is not one of these, and therefore no greater ordinal is (since L_{α_0} is transitive). Moreover, as shown above, $\alpha_0 \subset L_{\alpha_0}$.

As a consequence, $\Phi(x)$ takes the value α_0 in L_{α_0}, so $\alpha_0 \in \prod(L_{\alpha_0}) = L_{\alpha_0+1}$, which contradicts the definition of α_0. ∎

We can now proceed to check that all the axioms of $ZF+AF$ hold in the class L.

Axiom of Extensionality: If a is constructible, its elements are too. So this axiom holds.

Union Axiom: Let a be constructible, and α some ordinal for which $a \in L_\alpha$. As L_α is transitive, the set $b = \bigcup a$ is a subset of L_α. Since, moreover, the value in L_α of the expression $\exists y(y \,\varepsilon\, a \wedge x \,\varepsilon\, y)$ – which has one free variable x, and a single parameter a, in L_α by assumption – is clearly b, we can conclude that $b \in L_{\alpha+1}$, and that b is constructible.

Power-set Axiom: Let a be constructible, b the set of *constructible* subsets of a. The map $x \to \mathrm{od}(x)$ defined on b has some subset of the ordinals for its range, and some ordinal α will therefore be an upper bound to this subset. So for all $x \in b$ we will have $\mathrm{od}(x) \leqslant \alpha$, so that $b \subset L_\alpha$. Nothing prevents us from setting α so high that a is a member of L_α.

Take the expression $\forall u[u \,\varepsilon\, x \Rightarrow u \,\varepsilon\, a]$. It has one free variable, x, one parameter $a \in L_\alpha$. Its value in L_α is the set of elements x of L_α which are subsets of a; b then, by the definition of α. Thus $b \in L_{\alpha+1}$, and so b is constructible.

Scheme of Replacement: Let $R(x, y, a_0, \dots, a_{k-1})$ be a formula of two free variables whose parameters $a_0, \dots, a_{k-1}$ are in L; and suppose that, when interpreted in L, it defines a functional relation. Within the universe $\mathscr{U}$ this functional relation is defined by the formula $L(x) \wedge L(y) \wedge \wedge R^L(x, y, a_0, \dots, a_{k-1})$; both domain and range are subclasses of L.

Let b stand for the set of images of the elements of some constructible set a under this functional relation. We must show b to be constructible. Each element of it certainly is, so pick some ordinal α_0 exceeding the order of any element of b and big enough for $a, a_0, \dots, a_{k-1}$ to be in L_{α_0}. Then $b \subset L_{\alpha_0}$.

We now take the parameter-free $k+2$-ary functional relation $R(x, y, x_0, \ldots, x_{k-1})$, and apply to it the generalized reflection principle (p. 51), using the functional relation $y=L_\alpha$. This provides us with an ordinal $\beta>\alpha_0$ such that

$$\forall x \forall y \forall x_0 \ldots \forall x_{k-1} \big[x \in L_\beta \wedge y \in L_\beta \wedge x_0 \in L_\beta \wedge \ldots \wedge x_{k-1} \in L_\beta \rightarrow \big(R^L(x, y, x_0, \ldots, x_{k-1}) \leftrightarrow R^{L_\beta}(x, y, x_0, \ldots, x_{k-1})\big)\big].$$

Since $a_0, \ldots, a_{k-1} \in L_\beta$ we have

$$(*) \qquad \forall x \forall y \big[x \in L_\beta \wedge y \in L_\beta \rightarrow \big(R^L(x, y, a_0, \ldots, a_{k-1}) \leftrightarrow R^{L_\beta}(x, y, a_0, \ldots, a_{k-1})\big)\big].$$

Now the expression $\ulcorner R(u, v, a_0, \ldots, a_{k-1})\urcorner$ has two free variables u, v, and parameters $a_0, \ldots, a_{k-1} \in L_\beta$; call it $\Phi(u, v)$ for short. Thus Val (Φ, L_β) is the set of maps $f\colon \{u, v\} \rightarrow L_\beta$ such that $R^{L_\beta}(f(u), f(v), a_0, \ldots, a_{k-1})$. By (*), however, this set is the set of maps $f\colon \{u, v\} \rightarrow L_\beta$ for which $R^L(f(u), f(v), a_0, \ldots, a_{k-1})$. Consequently, $\mathrm{Val}(\mathbf{\exists} u(u \,\varepsilon\, a \wedge \Phi(u, v)), L_\beta)$ is the set of maps $f\colon \{v\} \rightarrow L_\beta$ for which $\exists x(x \in a \wedge R^L(x, f(v), a_0, \ldots, a_{k-1}))$; so $\mathrm{val}(\mathbf{\exists} u(u \,\varepsilon\, a \wedge \Phi(u, v)), L_\beta)$ is just the set of elements of L which are images of members of a under the functional relation $R^L(x, y, a_0, \ldots, a_{k-1})$; in fact, b, in view of the choice of β. Thus $b \in L_{\beta+1}$, and is constructible.

Axiom of Infinity: Every ordinal is constructible, as we have seen; so ω is constructible.

Axiom of Foundation: Let a be constructible and non-empty, and b be an element of it of lowest possible order. Every element of b has strictly lower order than b itself, so no element of b is an element of a. Thus $b \cap a=\emptyset$.

We have now proved that all the axioms of $ZF+AF$ hold in L.

Note that the ordinals of L are the ordinals of the universe $\mathscr{U}$. For all ordinals α are constructible, and clearly they all satisfy $On^L(\alpha)$. Conversely, a constructible α satisfying $On^L(\alpha)$ is transitive and linearly ordered by $\in$; so an ordinal.

THE AXIOM OF CONSTRUCTIBILITY HOLDS IN L

We must now show that the axiom of constructibility holds in L.

Absolute Formulas: A parameter-free formula $E(x_0, \ldots, x_{k-1})$ is called *absolute* if it can be obtained by applications of the following rules.

(1) A formula without quantifiers is absolute.

(2) If A and B are absolute, then $A \vee B$ and $A \wedge B$ are absolute.

(3) If $A(x, x_0, \ldots, x_{n-1})$ is absolute then $\exists x A(x, x_0, \ldots, x_{n-1})$ is absolute.

(4) If $A(x, y, x_0, \ldots, x_{n-1})$ is absolute, then

$$\forall x[x \in y \rightarrow A(x, y, x_0, \ldots, x_{n-1})],$$

a formula with free variables $y, x_0, \ldots, x_{n-1}$, is absolute.

On the basis of this definition we call a relation absolute if it can be defined in $\mathscr{U}$ by an absolute formula.

THEOREM: *Let $E(x_0, \ldots, x_{k-1})$ be an absolute formula, W a transitive class (that is, $\forall x \forall y[W(x) \wedge y \in x \rightarrow W(y)]$), and W' an extension of W (that is $\forall x(W(x) \rightarrow W'(x))$).* If *$a_0, \ldots, a_{k-1}$ are objects in W such that $E^W(a_0 \ldots, a_{k-1})$ is true, then $E^{W'}(a_0, \ldots, a_{k-1})$ is also true. Equivalently,*

$$\forall x_0 \ldots \forall x_{k-1}[W(x_0) \wedge \ldots \wedge W(x_{k-1}) \rightarrow (E^W(x_0, \ldots, x_{k-1}) \rightarrow E^{W'}(x_0, \ldots, x_{k-1}))].$$

PROOF: The proof is an informal induction on the length of the absolute formula E. For quantifier-free E the result is immediate since E^W and $E^{W'}$ are the same as E.

If E is $F \vee G$, take $a_0, \ldots, a_{k-1}$ in W; by the induction hypothesis

$$F^W(a_0, \ldots, a_{k-1}) \rightarrow F^{W'}(a_0, \ldots, a_{k-1})$$

and

$$G^W(a_0, \ldots, a_{k-1}) \rightarrow G^{W'}(a_0, \ldots, a_{k-1}),$$

so

$$(F^W(a_0, \ldots, a_{k-1}) \vee G^W(a_0, \ldots, a_{k-1})) \rightarrow (F^{W'}(a_0, \ldots, a_{k-1}) \vee G^{W'}(a_0, \ldots, a_{k-1})),$$

as required. The same proof works if E is $F \wedge G$.

If $E(x_0, \ldots, x_{k-1})$ is $\exists x F(x, x_0, \ldots, x_{k-1})$, take any $a_0, \ldots, a_{k-1}$ in W such that $E^W(a_0, \ldots, a_{k-1})$ holds; thus $\exists x[W(x) \wedge F^W(x, a_0 \ldots, a_{k-1})]$.

So for some a in W we have $F^W(a, a_0, \ldots, a_{k-1})$. By the induction hypothesis therefore $F^{W'}(a, a_0, \ldots, a_{k-1})$, whence $\exists x[W'(x) \wedge F^{W'}(x, a_0, \ldots, a_{k-1})]$. This is $E^{W'}(a_0, \ldots, a_{k-1})$.

Lastly, suppose that $E(y, x_0, \ldots, x_{k-1})$ is $\forall x[x \in y \rightarrow F(x, y, x_0, \ldots, x_{k-1})]$, and pick objects $b, a_0, \ldots, a_{k-1}$ in W such that $E^W(b, a_0, \ldots, a_{k-1})$. Then

$$\forall x[W(x) \rightarrow (x \in b \rightarrow F^W(x, b, a_0, \ldots, a_{k-1}))]$$

which is the same thing as

$$\forall x[x \in b \wedge W(x) \rightarrow F^W(x, b, a_0, \ldots, a_{k-1})].$$

But as b is in W, which is a transitive class, $x \in b \rightarrow W(x)$. Moreover, by the induction hypothesis, $(W(x) \wedge F^W(x, b, a_0, \ldots, a_{k-1})) \rightarrow F^{W'}(x, b, a_0, \ldots, a_{k-1})$. Thus the last inset formula yields

$$\forall x[x \in b \rightarrow F^{W'}(x, b, a_0, \ldots, a_{k-1})],$$

and so $E^{W'}(b, a_0, \ldots, a_{k-1})$. ∎

Note that the definition of *absolute relation* given here is equivalent to that of P. Cohen in [2]; it is not the same as K. Gödel's definition in [1].

If the relation $G(y, z_0, \ldots, z_{l-1})$ and the functional relation $y = F(x_0, \ldots, x_{k-1})$ are absolute, then the relation $G(F(x_0, \ldots, x_{k-1}), z_0, \ldots, z_{l-1})$ is absolute.

The relation in question can be defined by

$$\exists y[y = F(x_0, \ldots, x_{k-1}) \wedge G(y, z_0, \ldots, z_{l-1})],$$

so can be realized by application of rules (2) and (3) to the absolute relations given.

In particular, *if a is a set for which $y = a$ is equivalent to some absolute formula $R(y)$, and if $G(y, z_0, \ldots, z_{l-1})$ is absolute, then $G(a, z_0, \ldots, z_{l-1})$ is absolute.*

For the rest of this section we shall assume the axiom of foundation to hold for $\mathcal{U}$, and our efforts will be directed towards proving that *the functional relation $y = L_\alpha$ is absolute.* From this it will follow easily that the axiom of constructibility holds in L.

The (singulary) relations $On(x)$ and $x=\omega$ are absolute.

Nothing has to be proved for the former, since *AF* holds,[1] and so $On(x)$ can be written

$$\forall u \forall v[u \in x \wedge v \in x \to u \in v \vee u=v \vee v \in u] \wedge \forall u[u \in x \to \forall v(v \in u \to v \in x)].$$

For the latter we simply check the absoluteness of the following relations:
$x=\emptyset$, defined by $\forall y(y \in x \to y \notin x)$.
$y=x \cup \{x\}$, defined by $\forall z(z \in y \to z=x \vee z \in x) \wedge \forall z(z \in x \to z \in y) \wedge x \in y$.
$x=\omega$, defined by

$$On(x) \wedge \emptyset \in x \wedge \forall y[y \in x \to y \cup \{y\} \in x] \wedge \forall y[y \in x \to (y=\emptyset \vee \exists z(y=z \cup \{z\}))].$$

We have shown above that certain compositions of absolute relations are themselves absolute. Since the functional relation $y=L_\alpha$, whose absoluteness we are after, is defined by recursion, we shall want to show that functional relations defined recursively from absolute functional relations are themselves absolute. And we shall also want to show that the functional relation $y=\prod(x)$ is absolute. These results are the content of the next two lemmas.

LEMMA: *Let $y=H(x)$ be an absolute functional relation whose domain is the class of all maps defined on the ordinals. Then the functional relation F recursively defined for every ordinal α by*

$$F(\alpha)=H(F \restriction \alpha)$$

is absolute.

PROOF: We check successively the absoluteness of a dozen or more relations.
$z=\{x,y\}: x \in z \wedge y \in z \wedge \forall t(t \in z \to t=x \vee t=y)$.
$z=\langle x,y\rangle: z=\{\{x\},\{x,y\}\}$ (composition of absolute functional relations).
$y \subset x: \forall z[z \in y \to z \in x]$.
$z=x \cup y: x \subset z \wedge y \subset z \wedge \forall t(t \in z \to t \in x \vee t \in y)$.
$z=x \cap y: z \subset x \wedge z \subset y \wedge \forall t[t \in x \to (t \in y \to t \in z)]$.
$z=x \setminus y: z \subset x \wedge \forall t[t \in x \to (t \notin y \leftrightarrow t \in z)]$.
$z \subset x \times y: \forall t[t \in z \to \exists u \exists v[u \in x \wedge v \in y \wedge t=\langle u,v\rangle]]$.
$z \supset x \times y: \forall u[u \in x \to \forall v[v \in y \to \exists t(t \in z \wedge t=\langle u,v\rangle)]]$.

[1] The formula $On(\alpha)$ as written on p. 14 is patently not absolute. What is more, it can be shown not to be equivalent, in *ZF*, to any absolute formula.

$z = x \times y$: the conjunction of the previous two relations.
z maps x into y: $z \subset x \times y \wedge$

$$\forall u[u \in x \rightarrow \exists v \exists t(v \in y \wedge t \in z \wedge t = \langle u, v\rangle)]$$
$$\wedge \forall t \forall t' \forall u \forall v \forall v' [(t \in z \wedge t' \in z \wedge u \in x \wedge v \in y \wedge v' \in y$$
$$\wedge t = \langle u, v\rangle \wedge t' = \langle u, v'\rangle) \rightarrow v = v'].$$

f is a map with domain x: $\exists y(f$ maps x into $y)$.
f is a map: $\exists x(f$ is a map with domain $x)$.
$g = f \restriction x$ (a ternary relation):

$$\exists x'[x \subset x' \wedge (f \text{ is a map with domain } x') \wedge (g \text{ is a map with domain } x) \wedge g \subset f].$$

$y = f(x)$ (a binary functional relation): $(f$ is a map$) \wedge \langle x, y\rangle \in f$.

We can now write the formula $y = F(\alpha)$ which we are interested in as

$$On(\alpha) \wedge \exists f[(f \text{ is a map with domain } \alpha) \wedge$$
$$\forall \beta(\beta \in \alpha \rightarrow f(\beta) = H(f \restriction \beta)) \wedge y = H(f)].$$

If the relation $y = H(f)$ is absolute, so too is this. ▌

We shall need also the following absolute relations.

f is an injection of x into y (f is one-one):

$$(f \text{ maps } x \text{ into } y) \wedge \forall t \forall t' \forall u \forall u' \forall v[(t \in f \wedge t' \in f$$
$$\wedge u \in x \wedge u' \in x \wedge v \in y \wedge t = \langle u, v\rangle$$
$$\wedge t' = \langle u', v\rangle) \rightarrow u = u'].$$

f is a surjection from x on to y: $(f$ maps x into $y) \wedge$

$$\forall v[v \in y \rightarrow \exists u(u \in x \wedge \langle u, v\rangle \in f)].$$

$h = f \circ g$ (a binary functional relation):

$$\exists x \exists y \exists z [(g \text{ maps } x \text{ into } y) \wedge (f \text{ maps } y \text{ into } z)$$
$$\wedge (h \text{ maps } x \text{ into } z)]$$
$$\wedge \forall t[t \in h \rightarrow \exists u \exists v \exists w[\langle u, v\rangle \in g \wedge \langle v, w\rangle \in f$$
$$\wedge t = \langle u, w\rangle]].$$

Now the relation $y = L_\alpha$, whose absoluteness is in question, is a functional relation defined by recursion. We shall therefore apply the above lemma. For the relation $y = H(f)$ we take $y = \bigcup_{x \in dom(f)} \prod(f(x))$, and this we must show to be absolute ($dom(f)$ is the domain of f). It can be written, however, as

$$\forall x[x \in dom(f) \to \prod(f(x)) \subset y] \\ \wedge \forall z[z \in y \to \exists x[x \in dom(f) \wedge z \in \prod(f(x))]].$$

The following lengthy lemma will therefore complete the proof that $y = L_\alpha$ is absolute.

LEMMA: *The functional relation* $y = \prod(x)$ *is absolute.*

PROOF: As the relations $x = 0$ and $y = x \cup \{x\}$ are absolute, it follows that the relations $x = 1$, $x = 2$, ... are absolute (by successive composition). As $\sim$, $\mathbf{v}$, $\exists$, ε, $=$, and $\mathscr{V}$ are respectively 0, 1, 2, 3, 4, and the set of natural numbers $\geqslant 5$, the relations $x = \sim$, $x = \mathbf{v}$, $x = \exists$, $x = \varepsilon$, $x = =$, and $x = \mathscr{V}$ (this latter being written $x = \omega \setminus \{0, 1, 2, 3, 4\}$) are absolute. From this we can derive the absoluteness of several other relations, as follows.

$z = \mathscr{E}_0$ (the set of atomic expressions):

$$z = (\{\varepsilon\} \times \mathscr{V} \times \mathscr{V}) \cup (\{=\} \times \mathscr{V} \times \mathscr{V})$$

(a composition of absolute functional relations).

$z = \mathscr{E}_k$ (a functional relation of one argument, k):

$$\exists f[(f \text{ is a map with domain } \omega) \wedge f(0) = \mathscr{E}_0 \\ \wedge \forall n[n \in \omega \to f(n+1) = M(f(n))] \wedge z = f(k)],$$

where the functional relation $y = M(x)$ is given by the absolute formula

$$y = x \cup [\{\sim\} \times x] \cup [\{\mathbf{v}\} \times x \times x] \cup [(\{\exists\} \times \mathscr{V}) \times x].$$

$z \in \mathscr{E}$ (the set of parameter-free expressions): $\exists k(k \in \omega \wedge z \in \mathscr{E}_k)$.

$z = \mathscr{E}$: $\forall k[k \in \omega \to \mathscr{E}_k \subset z] \wedge \forall y[y \in z \to y \in \mathscr{E}]$.

$z \in \mathscr{E} \wedge y = w(z)$ (as before, $w(z)$ is the set of free variables of the expression z):

$$z \in \mathscr{E} \wedge \exists f[(f \text{ is a map with domain } \mathscr{E}) \\ \wedge \forall x \forall y[x \in \mathscr{V} \wedge y \in \mathscr{V} \to f(\langle \varepsilon, x, y \rangle) = f(\langle =, x, y \rangle) \\ = \{x, y\}] \wedge \forall x \forall y[x \in \mathscr{E} \wedge y \in \mathscr{E} \to (f(\langle \sim, x \rangle) = f(x) \\ \wedge f(\langle \mathbf{v}, x, y \rangle) = f(x) \cup f(y))] \wedge \forall x \forall y[x \in \mathscr{V} \wedge y \in \mathscr{E} \\ \to f(\langle \exists x, y \rangle) = f(y) \setminus \{x\}] \wedge y = f(z)].$$

We now have to show that the binary functional relation $y = {}^{w(F)}X$ is absolute. Note that the formula $y = {}^{Y}X$ (three variables, y, X, Y) is not absolute; for if Z is a transitive set, $(y = {}^{Y}X)^{Z}$ means that y is the set of maps from Y into X that are elements of Z, and clearly there is no reason for supposing that y is the set of *all* maps from Y into X. By the theorem on p. 86 then, the formula $y = {}^{Y}X$ is not absolute. Only because $w(F)$ is always finite can we expect to get the result we want.

Some more absolute relations first.

k is a natural number and $y = {}^{k}X$ (a functional relation of two arguments, X and k):

$$\exists f\,[(f \text{ is a map with domain } \omega) \wedge f(0) = \emptyset$$
$$\wedge\ \forall n\,[n \in \omega \rightarrow f(n+1) = N(n, X, f(n))] \wedge y = f(k)],$$

where the ternary functional relation $Z = N(n, X, Y)$ is given by the absolute formula

$$n \in \omega \wedge \forall g\,[g \in Z \rightarrow g \text{ maps } n+1 \text{ into } X]$$
$$\wedge\ \forall h \forall x\,[(h \in Y \wedge x \in X) \rightarrow h \cup \{\langle n, x\rangle\} \in Z].$$

${}^{w(F)}X \subset y$ (a ternary relation with arguments y, X, F):

$$\exists \varphi \exists n\,[n \in \omega \wedge (\varphi \text{ is a bijection from } w(F) \text{ on to } n)$$
$$\wedge\ \forall f\,(f \in {}^{n}X \rightarrow f \circ \varphi \in y)].$$

$y = {}^{w(F)}X$ (two arguments, X, F):

$${}^{w(F)}X \subset y \wedge \forall f\,[f \in y \rightarrow (f \text{ maps } w(F) \text{ into } X)].$$

y is a closed expression with parameters in X:

$$\exists F \exists \varphi\,[F \in \mathscr{E} \wedge (\varphi \text{ maps } w(F) \text{ into } X) \wedge y = \langle F, \varphi\rangle)].$$

$y = \mathscr{E}^{0}_{X}$ (y is the set of closed expressions with parameters in X):

$$\forall x\,[x \in y \rightarrow (x \text{ is a closed expression with parameters in } X)]$$
$$\wedge\ \forall F\,[F \in \mathscr{E} \rightarrow \forall \varphi\,[\varphi \in {}^{w(F)}X \rightarrow \langle F, \varphi\rangle \in y]].$$

$y \in \mathscr{E}^{x}_{X}$ (y is an expression with parameters in X and a single free variable x):

$$\exists F \exists \varphi\,[F \in \mathscr{E} \wedge x \in w(F)$$
$$\wedge\ (\varphi \text{ maps } w(F) \setminus \{x\} \text{ into } X) \wedge y = \langle F, \varphi\rangle].$$

$y = \mathscr{E}^{x}_{X}$: $\forall z\,[z \in y \rightarrow z \in \mathscr{E}^{x}_{X}]$

$$\wedge\ \forall F \forall \varphi\,[F \in \mathscr{E} \wedge x \in w(F) \wedge \varphi \in {}^{w(F)\setminus\{x\}}X \rightarrow \langle F, \varphi\rangle \in y].$$

$y \in \mathscr{E}_X^x$ and z is the closed expression obtained by substituting in y the element a of X for the variable x (five arguments, x, y, z, X, a):

$$\exists F \exists \varphi \exists \psi [F \in \mathscr{E} \wedge x \in w(F) \wedge (\varphi \text{ maps } w(F) \setminus \{x\} \text{ into } X)$$
$$\wedge (\psi \text{ maps } w(F) \text{ into } X) \wedge \varphi \subset \psi$$
$$\wedge \psi(x) = a \wedge y = \langle F, \varphi \rangle \wedge z = \langle F, \psi \rangle].$$

Φ is a closed expression with parameters in X and $\theta = \mathrm{Val}(\Phi, X)$ (this is a functional relation of two arguments, Φ, X; θ takes the values 0, 1):

$$\Phi \in \mathscr{E}_X^0 \wedge \exists f [(f \text{ maps } \mathscr{E}_X^0 \text{ into } \{0, 1\}) \wedge \forall a \forall b [a \in X \wedge b \in X$$
$$\rightarrow ((a \in b \wedge f(a \varepsilon b) = 1) \vee (a \notin b \wedge f(a \varepsilon b) = 0))]$$
$$\wedge \forall a \forall b [a \in X \wedge b \in X \rightarrow ((a = b \wedge f(a = b) = 1)$$
$$\vee (a \neq b \wedge f(a = b) = 0))] \wedge \forall \Psi [\Psi \in \mathscr{E}_X^0 \rightarrow f(\sim \Psi) = 1 - f(\Psi)]$$
$$\wedge \forall \Psi \forall \Psi' [\Psi \in \mathscr{E}_X^0 \wedge \Psi' \in \mathscr{E}_X^0 \rightarrow f(\Psi \vee \Psi') = f(\Psi) \cup f(\Psi')]$$
$$\wedge \forall x \forall \Psi [x \in \mathscr{V} \wedge \Psi \in \mathscr{E}_X^x \rightarrow ([\exists a (a \in X \wedge f(\Psi(a)) = 1)$$
$$\wedge f(\mathbf{\exists} x \Psi) = 1] \vee [\forall a (a \in X \rightarrow f(\Psi(a)) = 0) \wedge f(\mathbf{\exists} x \Psi) = 0])]$$
$$\wedge \theta = f(\Phi)].$$

$\Phi \in \mathscr{E}_X^x \wedge y = \mathrm{val}(\Phi, X)$ (y is the subset represented by the monadic expression Φ in the set X):

$$\forall a [a \in y \rightarrow (a \in X \wedge \mathrm{Val}(\Phi(a), X) = 1)]$$
$$\wedge \forall a [a \in X \rightarrow (\mathrm{Val}(\Phi(a), X) = 0 \vee a \in y)].$$

$y \in \prod(X)$ (two arguments, y, X):

$$\exists x \exists \Phi [x \in \mathscr{V} \wedge \Phi \in \mathscr{E}_X^x \wedge y = \mathrm{val}(\Phi, X)].$$

$y = \prod(X)$: $\forall z [z \in y \rightarrow z \in \prod(X)]$

$$\wedge \forall x \forall \Phi [x \in \mathscr{V} \wedge \Phi \in \mathscr{E}_X^x \rightarrow \exists z (z \in y \wedge z = \mathrm{val}(\Phi, X))].$$

In this way the relation $y = \prod(X)$ is shown to be absolute. ▮

THEOREM: *The relation $y = L_\alpha$ is absolute.*

It follows at once that the *axiom of constructibility holds in L*. For since $ZF + AF$ holds in the class L, the relation $(y = L_\alpha)^L$ is a functional relation. Moreover, according to the theorem on p. 86 we have

$$\forall y \forall \alpha [L(y) \wedge L(\alpha) \wedge (y = L_\alpha)^L \rightarrow y = L_\alpha],$$

since L is transitive and $y=L_\alpha$ is an absolute formula. So let a be some object in L. For some α_0 then, $a \in L_{\alpha_0}$; and since α_0 is an ordinal of L, there is a unique constructible set y_0 for which $(y_0=L_{\alpha_0})^L$ holds; as $(y_0=L_{\alpha_0})^L \rightarrow y_0=L_{\alpha_0}$, we get $y_0=L_{\alpha_0}$, so $a \in y_0$.

Thus $(a \in L_{\alpha_0})^L$ is true, and so $[\exists\alpha(a \in L_\alpha)]^L$ is also true.

The axiom of constructibility asserts that the three classes $\mathscr{U}$, V, L are identical (this axiom therefore implies AF). The axiom of constructibility is usually denoted $V=L$, which is an abbreviation of the formula $\forall x[V(x) \leftrightarrow L(x)]$ (when this notation is used, AF is understood to hold). What we have shown, therefore, is that if ZF is consistent, then so is $ZF+AF+V=L$.

The axiom of constructibility has many important consequences. We shall now show that it entails the axiom of choice, in particular – and even the choice principle stated on p. 69–, and the generalized continuum hypothesis. These two sentences are thus satisfied in L, and therefore consistent relative to the remaining axioms.

$V=L$ IMPLIES AC

Consider a universe $\mathscr{U}$ which satisfies the axiom of constructibility.

With a parameter-free formula we shall define on the class of well-orderings in $\mathscr{U}$ a functional relation $v=F(u)$ having the following property. If u is a well-ordering of the set X, $F(u)$ is a well-ordering of the set of monadic expressions with parameters in X.

Let $\mathscr{E}$ be the set of parameter-free expressions; it is a subset of V_ω, and we have already defined (on p. 41) an injection, itself without parameters, which maps $\mathscr{E}$ one-one into ω. This injection, K^{-1}, is the inverse of K, suitably restricted. Let j be the isomorphism from u on to its ordinal α.

A monadic expression with parameters in X is a pair $\langle E, \eta\rangle$ where $E \in \mathscr{E}$ and η maps some subset of $w(E)$ into X. If u well-orders X, each element of η, $\langle x, y\rangle$, where $x \in \mathscr{V}$ and $y \in X$, can be associated one-one with a pair of ordinals; namely x itself (remember the definition of $\mathscr{V}$ as $\omega\setminus 5$) and the 'place-number' of y in the well-ordering u of X (that is, the ordinal of the well-ordered initial segment $S_y(x, u)$). This pair of ordinals can itself be associated one-one with an ordinal, by the isomorphism from On^2 on to On, defined on p. 33. Thus η can be associated one-one with a finite sequence of ordinals – namely the sequence which takes in their natural order the ordinals associated with its elements. As a consequence, and using the injection K^{-1} noted above, we can find an injection from the set of monadic expressions with parameters in X into $\omega \times \sigma(On)$, a class which is well ordered by the well-ordering already defined on $\sigma(On)$. In this way we can obtain the well-ordering v of the set of monadic expressions with parameters in X.

We can now define recursively on the ordinals a parameter-free functional relation $y = B(\alpha)$ which to each ordinal α associates a well-ordering $B(\alpha)$ of L_α, such that for $\beta \leqslant \alpha$, L_β is an initial segment of L_α under $B(\alpha)$; and on L_β, $B(\alpha)$ and $B(\beta)$ are identical.

For the definition, suppose $B(\beta)$ suitably defined for all $\beta < \alpha$; then if α is a limit ordinal, $L_\alpha = \bigcup_{\beta<\alpha} L_\beta$ and so we write $B(\alpha) = \bigcup_{\beta<\alpha} B(\beta)$, the common extension of all the $B(\beta)$ for $\beta < \alpha$. If $\alpha = \beta + 1$, we use the fact that $B(\beta)$ well-orders L_β, so that $v_\beta = F(B(\beta))$ well-orders the set of monadic expressions whose parameters lie in L_β. For the well-ordering $B(\alpha)$ of $L_\alpha = \prod(L_\beta)$ we therefore write

$$\begin{aligned} x \leqslant y(\text{mod. } B(\alpha)) \leftrightarrow & [x \in L_\beta \wedge y \in L_\beta \wedge x \leqslant y(\text{mod. } B(\beta))] \\ & \vee [x \in L_\beta \wedge y \in L_\alpha \setminus L_\beta] \\ & \vee [x \in L_\alpha \setminus L_\beta \wedge y \in L_\alpha \setminus L_\beta \wedge \end{aligned}$$

the first expression defining x in L_β is earlier, in the well-ordering v_β of the monadic expressions with parameters in L_β, than the first expression defining y in L_β].

It is obvious that the ordering $B(\alpha)$ thus defined on L_α has all the required properties.

The functional relation $y = B(\alpha)$ is thus completely defined by recursion. A binary relation $R(x, y)$ without parameters which well-orders the whole of $\mathscr{U}$ is then available from the formula $\exists\alpha[On(\alpha) \wedge x \leqslant y\ (\text{mod. } B(\alpha))]$.

This shows that *the choice principle holds in* $\mathscr{U}$; by a result on p. 69, the axiom 'every set is ordinal definable' holds also in $\mathscr{U}$.

$V = L$ IMPLIES THE GENERALIZED CONTINUUM HYPOTHESIS

The proof of this result relies on a certain property of absolute relations, dealt with in the next theorem.

THEOREM: *Let* $\mathscr{U}$ *be a universe where* $ZF + AF + AC$ *holds, and* $E(x, y)$ *an* *absolute formula defining in $\mathscr{U}$ a functional relation $y=\Phi(x)$ of a single argument. Then $\overline{\overline{\Phi(x)}}\leqslant\overline{\overline{\mathscr{C}(x)}}+\aleph_0$ for every x in the domain of Φ.*

PROOF: Let a be in the domain of Φ, and α an ordinal for which $a\in V_\alpha$, $\Phi(a)\in V_\alpha$, and such that

$$\forall x\forall y\left[x\in V_\alpha\wedge y\in V_\alpha\rightarrow\left(E^{V_\alpha}(x,y)\leftrightarrow E(x\ \ y)\right)\right].$$

That there is such an ordinal is guaranteed by the reflection principle for the formula $E(x, y)$.

Let P stand for the transitive closure of $\{a\}$, that is $\mathscr{C}(\{a\})$. Since $\{a\}\subset V_\alpha$ and V_α is transitive, we have $P\subset V_\alpha$. Moreover, $P=\mathscr{C}(a)\cup\{a\}$, so $\overline{\overline{P}}=\overline{\overline{\mathscr{C}(a)}}+1$.

Let $\mathscr{G}$ be the set of all closed expressions that are satisfied in V_α and have all their parameters in P. By the Löwenheim/Skolem theorem (pp. 59–61) there is a subset X of V_α, including P, satisfying all the expressions in $\mathscr{G}$, and of cardinality not exceeding $\overline{\overline{P}}+\aleph_0$.

Since V_α is a transitive set, the axiom of extensionality holds in it; so the corresponding expression

$$\mathbf{\forall}x\mathbf{\forall}y\left[x=y\Leftrightarrow\mathbf{\forall}z(z\,\varepsilon\,x\Leftrightarrow z\,\varepsilon\,y)\right]$$

is in $\mathscr{G}$, and is therefore satisfied in X. Thus X is an extensional set.

By the corollary on p. 40 we can find an isomorphism j of X on to some transitive set Y; in consequence, Y will satisfy all the expressions in the set $\mathscr{G}'$ which is produced from $\mathscr{G}$ by replacing each parameter u (in P, and so in X) by $j(u)$.

But $j(u)=u$ for every $u\in P$, as can be seen by supposing u_0 to be an element of P for which $j(u_0)\neq u_0$, and one of least rank at that. For then $j(v)$ is equal to v for all $v\in u_0$; for $P=\mathscr{C}(\{a\})$ is transitive, so any such $v\in P$. However, $j(u_0)=\{j(v)\mid v\in u_0\}=u_0$, which contradicts the definition of u_0.

Thus $P\subset Y$, and Y satisfies all the expressions of $\mathscr{G}$; in addition, $\overline{\overline{Y}}=\overline{\overline{X}}\leqslant\overline{\overline{P}}+\aleph_0$.

We chose V_α to mirror the formula $E(x,y)$, so since a and $\Phi(a)$ are both in V_α, and $E(a,\Phi(a))$ is true, we have also $E^{V_\alpha}(a,\Phi(a))$. Thus the sentence $E(a,\Phi(a))$ is true in V_α, and so $\exists yE(a,y)$ is also true in V_α. The expression $\ulcorner\exists yE(a,y)\urcorner$ is thus satisfied in V_α, and has a single parameter

a, in P; so it belongs to $\mathscr{G}$, and is therefore satisfied in Y. Consequently, the formula $\exists y E(a, y)$ holds in Y, and for some $b \in Y$ we will have $E^Y(a, b)$.

But as $E(x, y)$ is an absolute formula and Y is a transitive set, we also have $E^Y(a, b) \to E(a, b)$. Thus $E(a, b)$, and so $b = \Phi(a)$. But b was in Y, so $\Phi(a) \in Y$, and Y being transitive, $\Phi(a) \subset Y$. Thus $\overline{\overline{\Phi(a)}} \leqslant \overline{\overline{Y}} \leqslant \overline{\overline{P}} + \aleph_0$, and so $\overline{\overline{\Phi(a)}} \leqslant \overline{\overline{\mathscr{C}(a)}} + \aleph_0$. ▮

As a particular case of this theorem, suppose a is an object of $\mathscr{U}$ defined by an absolute formula $E(y)$; in other words, that $\forall y[y = a \leftrightarrow E(y)]$; then $\bar{\bar{a}} \leqslant \aleph_0$.

It is enough for our purpose to put $x = \emptyset$ and to apply the above theorem to a formula $E(y)$ which does not contain x at all. We can deduce in this way that, for example, if $ZF + AF + AC$ holds in $\mathscr{U}$, then the formula $y = \mathscr{P}(\omega)$ is not equivalent to any absolute formula.

THEOREM: *Suppose again that $ZF + AF + AC$ holds in $\mathscr{U}$. Then $\overline{\overline{L_\alpha}} = \bar{\bar{\alpha}}$ for every ordinal $\alpha \geqslant \omega$. If a is a constructible set, $\overline{\overline{od(a)}} \leqslant \overline{\overline{\mathscr{C}(a)}} + \aleph_0$.*

PROOF: We have seen already that $\alpha \subset L_\alpha$, so that $\bar{\bar{\alpha}} \leqslant \overline{\overline{L_\alpha}}$. To show the converse inequality for all $\alpha \geqslant \omega$ we proceed by induction. For ω we note that $L_n = V_n$ for every $n \in \omega$ (a trivial induction proves this); thus $L_\omega = V_\omega$, so $\overline{\overline{L_\omega}} = \omega$.

Now $L_\alpha = \bigcup_{\beta < \alpha} \prod(L_\beta)$, and by the induction hypothesis $\overline{\overline{L_\beta}} \leqslant \bar{\bar{\beta}}$ if $\omega \leqslant \beta < \alpha$. But for infinite a we have $\overline{\overline{\prod(a)}} = \bar{\bar{a}}$, so

$$\overline{\overline{L_\alpha}} \leqslant \sum_{\beta < \alpha} \overline{\overline{\prod(L_\beta)}} = \sum_{\beta < \alpha} \overline{\overline{L_\beta}} \leqslant \sum_{\beta < \alpha} \bar{\bar{\beta}} \leqslant \bar{\bar{\alpha}} \cdot \bar{\bar{\alpha}} = \bar{\bar{\alpha}}.$$

An alternative proof of this would apply the previous theorem to the absolute relation $y = L_\alpha$. It is in this way that we demonstrate the second part of the theorem.

Recall that $\mathrm{od}(a)$ is the first ordinal α for which $a \in L_\alpha$; the relation $y = \mathrm{od}(x)$ is absolute since it can be written

$$On(y) \wedge x \in L_y \wedge \forall z[z \in y \to x \notin L_z],$$

and the relation $y = L_x$ is absolute. At once then, $\overline{\overline{\mathrm{od}(a)}} \leqslant \overline{\overline{\mathscr{C}(a)}} + \aleph_0$. ▮

With all this achieved, let $\mathscr{U}$ be a universe satisfying the axiom of constructibility. We have already proved that the axiom of choice holds in $\mathscr{U}$, so the theorems above can be applied. Any set a is constructible, so if $a \subset \aleph_\rho$ we have

$$\overline{\overline{\mathrm{od}(a)}} \leqslant \overline{\overline{\mathscr{C}(a)}} + \aleph_0 \leqslant \aleph_\rho,$$

since $\mathscr{C}(a) \subset \aleph_\rho$. Thus $\mathrm{od}(a) < \aleph_{\rho+1}$, so $a \in L_{\aleph_{\rho+1}}$. But a is any subset of $\aleph_\rho$, so $\mathscr{P}(\aleph_\rho) \subset L_{\aleph_{\rho+1}}$, whence $\overline{\overline{\mathscr{P}(\aleph_\rho)}} \leqslant \overline{\overline{L_{\aleph_{\rho+1}}}}$. Thus $\overline{\overline{\mathscr{P}(\aleph_\rho)}} \leqslant \aleph_{\rho+1}$. Cantor's theorem gives us the reverse inequality, so we can conclude that $\overline{\overline{\mathscr{P}(\aleph_\rho)}} = \aleph_{\rho+1}$, which is the generalized continuum hypothesis.

As far as arithmetical formulas are concerned, we have an analogue of the result obtained in Chapter VI.

If E is an arithmetical formula (a formula whose quantifiers are relativized to V_ω) that is derivable with the help of the axiom of choice and the generalized continuum hypothesis (and, more generally, the axiom of constructibility), then E is already derivable from ZF alone.

For $V_\omega = L_\omega$, as we have seen. A formula E with quantifiers relativized to L_ω is absolute, and so $E^L \to E$ is a consequence of *ZF*.

Let, then, $A_0, \ldots, A_{n-2}$, E be a proof of E from $ZF + V = L$. If A_i is an axiom, we know that A_i^L is a theorem of *ZF*. The sequence $A_0^L, \ldots, A_{n-2}^L$, E^L is thus a proof of E^L from *ZF* alone. With $E^L \to E$ also a theorem of *ZF*, we see that E itself is a theorem of this theory.

BIBLIOGRAPHY

The two standard works on this subject are

[1] Kurt Gödel, 1940, 'The Consistency of the Continuum Hypothesis', *Annals of Mathematics Studies* 3, Princeton University Press.

[2] Paul J. Cohen, 1966, *Set Theory and the Continuum Hypothesis*, W. A. Benjamin, New York.

An exposition of naive set theory can be found in

[3] Paul R. Halmos, 1960, *Naive Set Theory*, D. Van Nostrand, Princeton.

The fundamental ideas of elementary logic and model theory are discussed in detail in, for example,

[4] Elliott Mendelson, 1964, *Introduction to Mathematical Logic*, D. Van Nostrand, Princeton.

[5] G. Kreisel and J. L. Krivine, 1967, *Elements of Mathematical Logic (Model Theory)*, North Holland Publishing Company, Amsterdam.

[6] Roger C. Lyndon, 1966, *Notes on Logic*, D. Van Nostrand, Princeton.

[7] Joseph R. Shoenfield, 1967, *Mathematical Logic*, Addison-Wesley.
This last book contains (amongst much else) an important chapter on set theory.

On the subject of ordinal definable sets see

[8] John Myhill and Dana Scott, 1967, 'Ordinal Definability', Summer Institute on Axiomatic Set Theory, Los Angeles.

The proof given in the text of the generalized continuum hypothesis in L is close to that of

[9] Carol Karp, 1967, 'A Proof of the Relative Consistency of the Continuum Hypothesis', in *Sets, Models and Recursion Theory* (ed. by John N. Crossley), North-Holland Publishing Company, Amsterdam, pp. 1–32.

For the topic of absolute expressions see also

[10] Azriel Lévy, 1965, 'A Hierarchy of Formulas in Set Theory', *Memoirs of the American Mathematical Society* **57**.

SYNTHESE LIBRARY

Monographs on Epistemology, Logic, Methodology,
Philosophy of Science, Sociology of Science and of Knowledge, and on the
Mathematical Methods of Social and Behavioral Sciences

‡RISTO HILPINEN (ed.), *Deontic Logic: Introductory and Systematic Readings*. 1971, VII + 182 pp. Dfl. 45,—

‡EVERT W. BETH, *Aspect of Modern Logic*. 1970, XI + 176 pp. Dfl. 42,—

‡PAUL WEINGARTNER and GERHARD ZECHA, (eds.), *Induction, Physics, and Ethics. Proceedings and Discussions of the 1968 Salzburg Colloquium in the Philosophy of Science*. 1970, X + 382 pp. Dfl. 65,—

‡ROLF A. EBERLE, *Nominalistic Systems*. 1970, IX + 217 pp. Dfl. 42,—

‡JAAKKO HINTIKKA and PATRICK SUPPES, *Information and Inference*. X + 336 pp. Dfl. 60,—

‡KAREL LAMBERT, *Philosophical Problems in Logic. Some Recent Developments*. 1970, VII+176 pp. Dfl. 38,—

‡P. V. TAVANEC (ed.), *Problems of the Logic of Scientific Knowledge*. 1969, XII+ 429 pp. Dfl. 95,—

‡ROBERT S. COHEN and RAYMOND J. SEEGER (eds.), *Boston Studies in the Philosophy of Science*. Volume VI: *Ernst Mach: Physicist and Philosopher*. 1970, VIII+295 pp. Dfl. 38,—

‡MARSHALL SWAIN (ed.), *Induction, Acceptance, and Rational Belief*. 1970, VII+ 232 pp. Dfl. 40,—

‡NICHOLAS RESCHER *et al.* (eds.), *Essays in Honor of Carl G. Hempel. A Tribute on the Occasion of his Sixty-Fifth Birthday*. 1969, VII + 272 pp. Dfl. 46,—

‡PATRICK SUPPES, *Studies in the Methodology and Foundations of Science. Selected Papers from 1951 to 1969*. 1969, XII + 473 pp. Dfl. 70,—

‡JAAKKO HINTIKKA, *Models for Modalities. Selected Essays*. 1969, IX + 220 pp. Dfl. 34,—

‡D. DAVIDSON and J. HINTIKKA: (eds.), *Words and Objections: Essays on the Work of W. V. Quine*. 1969, VIII + 366 pp. Dfl. 48,—

‡J. W. DAVIS, D. J. HOCKNEY, and W. K. WILSON (eds.), *Philosophical Logic*. 1969, VIII + 277 pp. Dfl. 45,—

‡ROBERT S. COHEN and MARX W. WARTOFSKY (eds.), *Boston Studies in the Philosophy of Science*. Volume V: *Proceedings of the Boston Colloquium for the Philosophy of Science 1966/1968*. 1969, VIII + 482 pp. Dfl. 58,—

‡ROBERT S. COHEN and MARX W. WARTOFSKY (eds.), *Boston Studies in the Philosophy of Science*. Volume IV: *Proceedings of the Boston Colloquium for the Philosophy of Science 1966/1968*. 1969, VIII + 537 pp. Dfl. 69,—

‡NICHOLAS RESCHER, *Topics in Philosophical Logic*. 1968, XIV + 347 pp. Dfl. 62,—

‡GÜNTHER PATZIG, *Aristotle's Theory of the Syllogism. A Logical-Philological Study of Book A of the Prior Analytics*. 1968, XVII + 215 pp. Dfl. 45,—

‡C. D. BROAD, *Induction, Probability, and Causation. Selected Papers*. 1968, XI + 296 pp. Dfl. 48,—

‡ROBERT S. COHEN and MARX W. WARTOFSKY (eds.), *Boston Studies in the Philosophy of Science*. Volume III: *Proceedings of the Boston Colloquium for the Philosophy of Science 1964/1966*. 1967, XLIX + 489 pp. Dfl. 65,—

‡GUIDO KÜNG, *Ontology and the Logistic Analysis of Language. An Enquiry into the Contemporary Views on Universals*. 1967, XI + 210 pp. Dfl. 38,—

*EVERT W. BETH and JEAN PIAGET, *Mathematical Epistemology and Psychology*. 1966. XXII + 326 pp. Dfl. 58,—

*EVERT W. BETH, *Mathematical Thought. An Introduction to the Philosophy of Mathematics*. 1965, XII + 208 pp. Dfl. 32,—

‡PAUL LORENZEN, *Formal Logic*. 1965, VIII + 123 pp. Dfl. 22,—

‡GEORGES GURVITCH, *The Spectrum of Social Time*. 1964, XXVI + 152 pp. Dfl. 20,—

‡A. A. ZINOV'EV, *Philosophical Problems of Many-Valued Logic*. 1963, XIV + 155 pp. Dfl. 28,—

‡MARX W. WARTOFSKY (ed.), *Boston Studies in the Philosophy of Science*. Volume I: *Proceedings of the Boston Colloquium for the Philosophy of Science, 1961–1962*. 1963, VIII + 212 pp. Dfl. 22,50

‡B. H. KAZEMIER and D. VUYSJE (eds.), *Logic and Language. Studies dedicated to Professor Rudolf Carnap on the Occasion of his Seventieth Birthday*. 1962, VI + 246 pp. Dfl. 32,50

*EVERT W. BETH, *Formal Methods. An Introduction to Symbolic Logic and to the Study of Effective Operations in Arithmetic and Logic*. 1962, XIV + 170 pp. Dfl. 30,—

*HANS FREUDENTHAL (ed.), *The Concept and the Role of the Model in Mathematics and Natural and Social Sciences. Proceedings of a Colloquium held at Utrecht, The Netherlands, January 1960*. 1961, VI + 194 pp. Dfl. 30,—

‡P. L. R. GUIRAUD, *Problèmes et méthodes de la statistique linguistique*. 1960, VI + 146 pp. Dfl. 22,50

*J. M. BOCHEŃSKI, *A Precis of Mathematical Logic*. 1959, X + 100 pp. Dfl. 20,—

Sole Distributors in the U.S.A. and Canada:

*GORDON & BREACH, INC., 150 Fifth Avenue, New York, N.Y. 10011
‡HUMANITIES PRESS, INC., 303 Park Avenue South, New York, N.Y. 10010

10/22/71
m8h

AN